FASTtrack

Pharmaceutical Compounding and Dispensing

FASTtrack

Pharmaceutical Compounding and Dispensing

Christopher A Langley
Professor of Pharmacy Law and Practice
Aston University School of Pharmacy
Birmingham, UK

Dawn Belcher
Teaching Fellow, Pharmacy Practice
Aston University School of Pharmacy
Birmingham, UK

Pharmaceutical Press
London

Published by Pharmaceutical Press

66-68 East Smithfield, London E1W 1AW, UK

© Royal Pharmaceutical Society of Great Britain 2012

(PP) is a trade mark of Pharmaceutical Press

Pharmaceutical Press is the publishing division of the Royal Pharmaceutical Society

First edition published 2008.
Second edition published 2012
Reprinted 2014

Typeset by Newgen Imaging Systems, India
Printed in Great Britain by TJ International, Padstow, Cornwall

ISBN 978 0 85711 055 8

A catalogue record for this book is available from the British Library.

Related videos can be found at www.pharmpress.com/PCDvideos
Please enter the access code PCD2edOV.

Contents

Introduction to the *FASTtrack* series

FASTtrack is a new series of revision guides created for undergraduate pharmacy students. The books are intended for use together with textbooks and reference books as an aid to revision to help guide students through their exams. They provide essential information required in each particular subject area. The books are also useful for pre-registration trainees preparing for the General Pharmaceutical Council's (GPhC) registration examination, and to practising pharmacists as a quick reference text.

The content of each title focuses on what pharmacy students really need to know in order to pass exams. Features include*:
- concise bulleted information
- key points
- tips for the student
- multiple choice questions (MCQs) and worked examples
- case studies
- simple diagrams.

The titles in the *FASTtrack* series reflect the full spectrum of modules for the undergraduate pharmacy degree.

Titles include*:
Applied Pharmaceutical Practice
Complementary and Alternative Medicine
Law and Ethics in Pharmacy Practice
Managing Symptoms in the Pharmacy
Pharmaceutical Compounding and Dispensing
Pharmaceutics: Dosage form and design
Pharmaceutics: Drug delivery and targeting
Pharmacology
Physical Pharmacy (based on Florence and Attwood's *Physicochemical Principles of Pharmacy*)
Therapeutics

There is also an accompanying website that includes extra MCQs, further title information and sample content: www.fasttrackpharmacy.com.

If you have any feedback regarding this series, please contact us at feedback@fasttrackpharmacy.com.

* Note: not all features are in every title in the series.

Preface

This book has been written as a student guide to extemporaneous pharmaceutical compounding and dispensing. It has been designed to assist the student compounder in understanding the key dosage forms encountered within extemporaneous dispensing.

Included is a summary of the key principles relating to labelling and packaging, along with a summary of the formulation of each dosage type. In addition, worked examples and questions have been included to allow the compounder to practise extemporaneous formulation exercises.

Christopher A Langley
Dawn Belcher
February 2012

About the authors

CHRISTOPHER A LANGLEY is a qualified pharmacist who graduated from Aston University in 1996 and then undertook his pre-registration training at St Peter's Hospital in Chertsey. Upon registration, he returned to Aston University to undertake a PhD within the Medicinal Chemistry Research Group before moving over full-time to Pharmacy Practice. He is currently employed as Professor of Pharmacy Law and Practice, specialising in teaching the professional and legal aspects of the degree programme.

His research interests predominantly surround pharmacy education but he is also involved in research examining the role of the pharmacist within both primary and secondary care. This includes examining the pharmacist's role in public health and the reasons behind and possible solutions to the generation of waste medication.

DAWN BELCHER is a qualified pharmacist who graduated from the Welsh School of Pharmacy in 1977 and then undertook her pre-registration training with Boots the Chemist at their Wolverhampton store. After registration she worked as a relief manager and later as a pharmacy manager for Boots the Chemist until 1984. While raising a family she undertook locum duties for Boots the Chemist and in 1986 became an independent locum working for a small chain of pharmacies in the West Midlands while also working for Lloyds Chemist.

In 1989 she began sessional teaching with the Pharmacy Practice group at Aston University which continued until she took a permanent post in 2001. She now enjoys teaching practical aspects of pharmacy practice while still keeping an association with Lloydspharmacy where she is employed as a relief manager.

Online material

The following is a list of the videos available on www.pharmpress.com/PCDvideos. Please enter access code PCD2edOV.
1. Dispensing Solutions (5 minutes, 7 seconds)
2. Dispensing Suspensions (8 minutes, 17 seconds)
3. Dispensing Emulsions (4 minutes, 11 seconds)
4. Dispensing Creams (5 minutes, 13 seconds)
5. Dispensing Ointments (8 minutes, 6 seconds)
6. Dispensing Suppositories (6 minutes, 5 seconds)
7. Dispensing Powders (9 minutes, 35 seconds)

chapter 1
Introduction

Layout of this text

Accurate and effective pharmaceutical formulation is a key skill which must be mastered by all student pharmacists and pharmaceutical technicians. This book is intended to be a guide to assist the student compounder in practising exercises relating to the key dosage forms encountered within extemporaneous dispensing. This revised edition includes web-based videos demonstrating the preparation of the different types of products. Look out for the symbol and follow the web link www.pharmpress.com/PCDvideos.

The book has been designed as a stand-alone revision text and summarises the key points behind the manufacture of common extemporaneous dosage forms, along with a series of worked examples and questions (with answers) for students to use for self-learning.

Each chapter is set out as follows:

- A **chapter overview** box summarising the key points covered in the chapter
- An **introduction and overview** of the product type
- A **general method** for the preparation of the product type
- A collection of **worked examples** to aid understanding and to include details on suitable labelling and packaging
- A series of **self-assessment questions** which it is expected that the student would work through independently. The answers to the questions can be found at the end of the book.

When a prescription is received for an extemporaneous product there are a number of considerations to be made prior to dispensing. Within each chapter, the worked examples section will take a number of different preparations and expand on their compounding using the following subheadings:

1. **Use of the product**
2. **Is it safe and suitable for the intended purpose?**
3. **Calculation of formula for preparation**
4. **Method of preparation**
   a. Solubility where applicable
   b. Vehicle/diluent
   c. Preservative
   d. Flavouring when appropriate
5. **Choice of container**
6. **Labelling considerations**
   a. Title

    b. Quantitative particulars

    c. Product-specific cautions (or additional labelling requirements)

    d. Directions to patient – interpretation of Latin abbreviations where necessary

    e. Recommended *British National Formulary* cautions when suitable

    f. Discard date

    g. Sample label (you can assume that the name and address of the pharmacy and the words 'Keep out of the reach and sight of children' are pre-printed on the label)

7. Advice to patient

In all the worked examples, the information provided in this text has been fully referenced. Wherever possible, the following reference texts have been used:

- *British Pharmacopoeia* (2007, London: TSO).
- *British National Formulary*, 61st edn (2011, London: BMJ Group and Pharmaceutical Press).
- *Martindale, The Complete Drug Reference*, 35th edn (London: Pharmaceutical Press).

For some information (e.g. solubility data) it has been necessary to use older reference sources. Where this has happened, details of the references used have been fully annotated within the text. However, it should always be remembered that, wherever possible, compounders should use the most up-to-date reference source available.

In addition to the product-type chapters, this chapter contains a summary of the key storage, labelling and packaging requirements for extemporaneous dosage forms.

Overview

Upon completion of this chapter, you should be able to:

- understand the key principles behind labelling of pharmaceutical preparations including:
- How to label products for both internal and external use.
- The importance of auxiliary labels.
- The rationale behind choosing an appropriate discard date.
- identify the different pharmaceutical packaging available
- select appropriate packaging for different pharmaceutical formulations.

Storage and labelling requirements

Storage

All products dispensed extemporaneously require some form of additional storage instructions to be detailed on the label. This information can be the addition of just a product expiry date through to a number of important additional auxiliary labels.

The summary list given in Table 1.1 is to be used as a guide in the absence of any guidance from the official pharmaceutical texts. It should be remembered that the information in this table is to be used only as a guide. Any information on additional labelling or expiry dates in the official texts will supersede the information provided in Table 1.1.

All suggested expiry dates included in Table 1.1 and other sections of this book are to be used as a guide only and are based on historical practice.

General principles of labelling

Every extemporaneously prepared preparation will require a label to be produced before the product can be dispensed or sold to the patient. The accuracy of the label is paramount as it conveys essential information to the patient on the use of the preparation.

Although the pharmacist or other healthcare practitioner may counsel patients when the medication is handed over, it is unlikely that patients will be able to remember all the information that they are given verbally. The label therefore acts as a permanent reminder of the key points that patients need to know.

Tips

Nowadays it is common practice to assign a maximum of a 2-week discard to any extemporaneously prepared product. Consideration should always be given to assigning a shorter discard date.

KeyPoints

The label of a pharmaceutical product has many functions:
- To indicate clearly the contents of the container
- To indicate clearly to patients how and when the medicinal product should be taken or used
- To indicate clearly to patients how the product should be stored and for how long
- To indicate clearly to patients any warnings or cautions of which they need to be made aware.

Table 1.1 A guide to auxiliary labels and discard dates for extemporaneous preparations.

Preparation	Container	Important auxiliary labels	Suggested discard date
Applications	Amber fluted bottle with CRC	For external use only	4 weeks
Capsules	Amber tablet bottle with CRC	See *BNF* for advisory labels recommended for active ingredient	3 months
Creams and Gels	Amber glass jar or collapsible metal tube	For external use only	4 weeks

(continued)

Preparation	Container	Important auxiliary labels	Suggested discard date
Dusting Powders	Plastic jar preferably with a perforated, reclosable lid	For external use only Not to be applied to open wounds or raw weeping surfaces Store in a dry place	3 months
Ear Drops	Hexagonal amber fluted glass bottle with a rubber teat and dropper closure	For external use only	4 weeks
Elixirs	Plain amber medicine bottle with CRC		4 weeks
Emulsions	Plain amber medicine bottle with CRC	Shake the bottle	4 weeks
Enemas	Amber fluted bottle with CRC	For rectal use only* Warm to body temperature before use	4 weeks
Gargles and Mouthwashes	Amber fluted bottle with CRC	Not to be taken* Do not swallow in large amounts	4 weeks
Inhalations	Amber fluted bottle with CRC	Not to be taken* Shake the bottle	4 weeks
Linctuses	Plain amber medicine bottle with CRC		4 weeks
Liniments and Lotions	Amber fluted bottle with CRC	For external use only Shake the bottle Avoid broken skin	4 weeks
Mixtures and Suspensions	Plain amber medicine bottle with CRC	Shake the bottle	4 weeks
Nasal Drops	Hexagonal amber fluted glass bottle with a rubber teat and dropper closure	Not to be taken*	4 weeks
Ointments	Amber glass jar	For external use only	3 months
Pastes	Amber glass jar	For external use only	3 months
Pessaries	Wrapped in foil and packed in an amber glass jar	For vaginal use only*	3 months
Powders (individual)	Wrapped in powder papers and packed in a cardboard carton	Store in a dry place Dissolve or mix with water before taking See *BNF* for advisory labels recommended for active ingredient	3 months
Suppositories	Wrapped in foil and packed in an amber glass jar	For rectal use only* See *BNF* for advisory labels recommended for active ingredient	3 months

BNF – British National Formulary; CRC – Child Resistant Closure
* See General principles of labelling below

Appearance
Correct position
– **Medicine bottles:** The label should be on the front of a medicine bottle about a third of the way down the container. The front of an internal bottle is the curved side and the front of a fluted bottle is the plain side.
– **Cartons:** The label should be placed on the large side of the carton. If there is not enough room on a single side of the carton for the entire label, it should be placed around the carton, ensuring that all the information is visible.
– **Ointment jars:** The label should be placed on the side of the jar, ensuring that the contents of the label are visible when the top is placed on the jar.
– Ensure that the patient can open the container without destroying the label (e.g. when labelling cartons).
– Ensure the label is positioned with care and is straight, not crooked.

Clean
– Ensure the container is clean before packing the product, then clean the outside before affixing the label. Never pour any liquids into a pre-labelled container as this risks spoiling the label with drips of the medicament.

Secure
– Ensure that the label is secure before dispensing the product to the patient. The main reason for labels not sticking to product containers is because of a dirty or greasy container.

Information
Legible
– Always check label print size and quality to ensure that it can be read clearly. If there is too much information to place on one label, consider placing the additional information on a secondary label, rather than reducing the size of the print or trying to include too much information on one label.

Concise
– Although it is important that sufficient information is placed on the label, it must be remembered that it is important not to confuse the patient by placing too much information on the label. If the label contains too much information, rather than assisting patients, they may feel overwhelmed and as a result they may read none of the information.

Adequate
– Ensure that sufficient information is given. For example, the term 'when required' raises the questions how much? How often? When required for what?

Intelligible

– The wording of the information on the label must be in plain English, be easily understandable and use unambiguous terms. It must always be remembered that patients may feel embarrassed to ask for further clarification on the meaning of complicated words used on the label.

Accurate

– It is important that the title is accurate, the instructions are accurate and the patient name is complete and accurate.

Dispensed type labels

In the UK, detailed requirements for labelling of medicinal products are contained in the Medicines Act 1968 and in amendments to that Act made by Statutory Instrument. The legislation distinguishes between labelling of a medicinal product for sale and labelling for a dispensed product when lesser requirements apply.

1. All labels for dispensed medicines must have the name of the patient – preferably the full name, not just initials – and if possible the title of the patient (Mr, Mrs, Miss, Master, Ms, etc.) as this helps to distinguish between family members. The date and the name and address of the pharmacy are also legally required. This will normally automatically appear on most computer labelling systems with the date being reset automatically. The words 'Keep out of the reach and sight of children' are also legally required, but most labels used for dispensing purposes are already pre-printed with these words.

2. All labels must state the name of the product dispensed, the strength where appropriate, and the quantity dispensed.

3a. Products for internal use

   The title of an extemporaneous preparation should be given if it is an official product (i.e. one with an accepted formula that can be found in an official text). The title should be as quoted in the official text (for example, 'Ammonia and Ipecacuanha Mixture BP').

   If it is an unofficial product (that is, a product made from an individual formula, for example, a doctor's own formula) it may be labelled 'The Mixture' or 'The Solution', etc. Unofficial products must state the full quantitative particulars on the label (i.e. the formula must be stated on the label). For preparations intended for internal use, this is expressed as the amount of ingredient per unit dose.

KeyPoints

Remember, the label of a pharmaceutical product must be in the right place and contain the right information. The following need to be taken into consideration:

Appearance
- Correct position
- Clean
- Secure

Information
- Legible
- Concise
- Adequate
- Intelligible
- Accurate

For example, the quantitative particulars for a Sodium Chloride BP solution 4% with a dose of 10 mL bd could be labelled as:

Table 1.2 The Solution

Each 10 mL dose contains:

Sodium Chloride BP	400 mg
Freshly boiled and cooled purified water	to 10 mL

3b. Products for external use

Labels for preparations not intended for oral use require slightly different labelling. If the product being made is official, the official title should be used (e.g. 'Sodium Bicarbonate Ear Drops BP' or 'Sodium Chloride Mouthwash BP').

If the product is an unofficial product the label title may reflect the type of external product:

e.g. 'The Nose Drops'
    'The Ear Drops'
    'The Mouthwash'
    'The Lotion'
    'The Enema', etc.

As with preparations intended for oral use, unofficial products for external use need to be labelled with the full quantitative particulars. In the case of products for external use, the quantitative particulars are expressed as the complete formula.

Therefore, the quantitative particulars for 100 mL Sodium Chloride BP Lotion 4% would be labelled:

Table 1.3 The Lotion

Containing:

Sodium Chloride BP	4 g
Freshly boiled and cooled purified water	to 100 mL

or

Table 1.4 The Lotion

Containing:

Sodium Chloride BP	4%
Freshly boiled and cooled purified water	to 100%

Similarly, the quantitative particulars for 200 mL of Sodium Chloride BP Lotion 4% would be labelled:

Table 1.5 The Lotion

Containing:

Sodium Chloride BP	8 g
Freshly boiled and cooled purified water	to 200 mL

or

Table 1.6 The Lotion

Containing:

Sodium Chloride BP	4%
Freshly boiled and cooled purified water	to 100%

4. Labels must also include an expiry date. See Table 1.1 for guidance. The Medicines Act 1968 (as amended) requires medicinal products to specify a month and year after which the product should not be used. However, in practice this can cause confusion and an alternative format is to show expiry as a single discard date: for example, 'Discard after 31.01.12'.

5. Warning labels may also be required. These may be pharmaceutical or pharmacological warnings (see labelling appendix in the *British National Formulary*). Generally if there is a choice between two warning labels with equivalent meaning, the positive one should be chosen (e.g. 'For rectal use only' is preferable to 'Do not swallow' for suppositories).

Table 1.1 gives guidance on the use of additional auxiliary labels. Within the UK, the term 'For external use only' is used on any preparation intended for external use. The Medicines Act 1968 defines products for external use as embrocations, liniments, lotions, liquid antiseptics, other liquids or gels for external use.

However, traditionally, for the following dosage forms, alternative labels have been employed instead of 'For external use only' to reflect more closely the intended purpose of the product. These alternative labels are (indicated on Table 1.1 by '*'):

- Enemas: 'For rectal use only'
- Gargles and mouthwashes: 'Not to be taken'
- Inhalations: 'Not to be taken'
- Nasal drops: 'Not to be taken'
- Pessaries: 'For vaginal use only'
- Suppositories: 'For rectal use only'.

Pharmacists should use their professional judgement when deciding which auxiliary labels should be applied to different

KeyPoints

Remember, the label of a pharmaceutical product will need to indicate the contents.

For official preparations, it will be sufficient to put the official title (as this indicates the contents).

For unofficial products, the formula will need to be detailed on the label. For internal products this information is given per dose; for external products, per container.

pharmaceutical dosage forms. As it is accepted practice within the UK to use the terms outlined above, these will be the terms used within each of the product chapters.

6. All directions on labels should use active rather than passive verbs. For example, 'Take two' (not 'Two to be taken'), 'Use one' (not 'One to be used'), 'Insert one' (not 'One to be inserted'), etc.

7. Where possible, adjacent numbers should be separated by the formulation name. For example, 'Take two three times a day' could allow for easy misinterpretation by the patient. Therefore, ideally, the wording on this label would include the formulation, e.g. 'Take two tablets three times a day'. The frequency and quantity of individual doses are always expressed as words rather than numerals (i.e. 'two' not '2').

8. Liquid preparations for internal use usually have their dose expressed as a certain number of 5 mL doses. This is because a 5 mL spoon is the normal unit provided to patients to measure their dose from the dispensed bottle. Therefore if a prescription called for the dosage instruction 10 mL tds, this would be expressed as 'Take two 5 mL spoonfuls three times a day'. Paediatric prescriptions may ask for a 2.5 mL dose: in this case, the label would read 'Give a 2.5 mL dose using the oral syringe provided'. Note here the use of the word 'Give', as the preparation is for a child and would be given to the patient by the parent or guardian.

9. Remember the label on a medicine is included so that the item can be identified, and the patient instructed as to the directions for use. Therefore, simple language should always be used.

■ Never use the word 'Take' on a preparation that is not intended for the oral route of administration.

■ Use 'Give' as a dosage instruction on products for children as a responsible adult should administer them.

■ Only use numerals when quoting the number of millilitres to be given or taken. All other dosage instructions should use words in preference to numerals.

■ Always be prepared to give the patient a verbal explanation of the label.

Pharmaceutical packaging

All dispensed medicinal products will need to be dispensed to the patient in a suitable

KeyPoints

Remember, labels for extemporaneous products need to contain the following:
■ Full name (including title) of the patient
■ Name of the product
■ Quantitative particulars (for unofficial products)
■ Appropriate expiry date
■ Additional warnings (where appropriate)
■ Directions for use

These are in addition to the name and address of the pharmacy and the words 'Keep out of the reach and sight of children'.

product container. The function of a container for a medicinal product is to maintain the quality, safety and stability of its contents.

Although different pharmaceutical preparations will be packaged in different containers depending on the product type, pharmaceutical packaging can largely be grouped into a few main types.

KeyPoints

The ideal container should be:
- Robust enough to protect the contents against crushing during handling and transport
- Convenient to use in order to promote good patient compliance (i.e. encourage patients to take their medication at the correct times)
- Easy to open and close, especially if the medication is for an elderly or arthritic patient
- Constructed of materials which do not react with the medicine, so the materials of construction should be inert
- Sufficiently transparent to allow for inspection of the contents in the case of liquid preparations.

Tablet bottles

Tablet bottles come in a variety of shapes and sizes and are usually made of either glass or plastic (Figure 1.1). Generally, tablet bottles are coloured amber to reduce the likelihood of the contents reacting with light. They are used for solid, single-dose preparations that are intended for oral use (i.e. tablets and capsules).

In normal circumstances, all tablet bottles would be fitted with child-resistant closures. Although not child-proof, these closures reduce the possibility of access to medication by children. There are a number of different types of child-resistant closures on the market. Consideration should be given to the patient when using child-resistant closures, as some patient groups (e.g. the elderly and arthritic patients) may not be able to open the container to access their medication. This can lead to non-compliance or reduced compliance.

Figure 1.1.
A selection of tablet bottles.

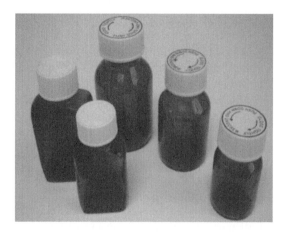

Medical bottles

Plain amber medicine bottles

Plain amber medicine bottles can be used to package all internal liquid preparations. Traditional amber medicine bottles used in the UK have two different sides, one curved and one flat (Figure 1.2). The label or labels are usually placed on the curved side of the bottle as the patient's natural action will be to pick the bottle up with the curved side of the bottle facing the inside of the palm. By placing the label on the curved side, this will mean the label is on the upper side of the bottle when in use. This will prevent the label becoming damaged by any dribbles of liquid running down the side of the bottle during pouring of a dose.

Figure 1.2.
A selection of plain amber medicine bottles.

Plain amber medicine bottles come in a variety of sizes. The capacity of each bottle is traditionally marked on the bottom of the container in millilitres. In the UK, plain amber medicine bottles come in the following sizes: 50 mL, 100 mL, 150 mL, 200 mL, 300 mL and 500 mL.

As with tablet bottles (see above), chid-resistant closures should be employed whenever possible.

Fluted amber medicine bottles

Fluted (or ribbed) amber medicine bottles are similar to the plain amber medicine bottles but, instead of having a flat plain side, this side is curved and contains a number of ridges or grooves running from the top of the bottle down to the bottom (Figure 1.3). The ridges or grooves are intended to be both a visual and a tactile warning to the patient or carer that the contents of the bottle are not to be administered via the oral route (the tactile nature of the warning is particularly useful for blind or partially sighted patients). For this reason, these types of container are often referred to as 'external medicine bottles' or 'poison bottles'. There is a legal requirement (Medicines Act 1968) in the UK that fluted bottles be used with specific types of pharmaceutical preparations: embrocations, liniments, lotions, liquid antiseptics or other liquids for external use.

Figure 1.3.
A selection of fluted amber medicine bottles.

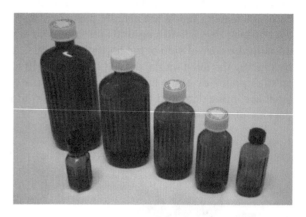

As with plain amber medicine bottles, the label is placed on the smooth curved side of the bottle and the capacity of each bottle is traditionally marked on the bottom of the container in millilitres. In the UK, fluted amber medicine bottles typically come in the following sizes: 50 mL, 100 mL and 200 mL, although other sizes may be available. Fluted dropper bottles are also available.

As with tablet bottles (see above), child-resistant closures should be employed whenever possible.

Calibrated containers for liquid preparations

Liquid preparations are normally made up to volume in a conical measure (Figure 1.4). There are occasions where a tared or calibrated bottle may be used. A tared bottle is normally only employed when, because of the viscosity of the final product, the transference loss from the measure to the container would be unacceptable. For example, Kaolin Mixture BP is a very dense suspension and transference may cause problems; similarly, a thick emulsion will also prove difficult and time-consuming to transfer in its entirety because of the viscosity of the finished product.

Points to consider when taring a container:

- The volume of water added to the container to be tared must be identical to that of the product being prepared and must be accurately measured using a conical measure.
- When poured into the container the meniscus is marked. (A simple method is to use a small adhesive label to mark the position and thus produce a measure with just one graduation.)
- The water is removed from the bottle and the bottle drained.

Figure 1.4.
A selection of conical measures.

- The prepared mixture is transferred to the calibrated bottle, the measure or mortar used in the preparation of the product is rinsed with more vehicle and this is added to the bottle.
- Any liquid ingredients are added and the mixture is made up to volume using the vehicle.
- Remove the meniscus marker before dispensing the preparation to the patient.

Please note: unless the bottle is thoroughly dried after taring, this method can only be used where water is one of the ingredients of the mixture as putting medicine into a wet bottle is considered to be bad practice.

Cartons

Cardboard cartons come in a variety of differing sizes, the sizes being dependent on the manufacturer (Figure 1.5). They tend to be rectangular in shape and the label is placed on the larger side of the box. They are used to package blister strips of tablets or capsules, powder papers and other pharmaceutical products that may be of a shape that is not suitable for labelling. Although it is good dispensing practice to label the primary container of a medicinal product, in some cases this isn't possible. By placing the primary container into a labelled carton, this provides the next best method for labelling the product in question (for example, the labelling of very small eye/ear/nose dropper bottles). Additional care must be exercised in the storage of pharmaceutical products in cardboard cartons as they do not come with child-resistant closures.

Figure 1.5.
A selection of cartons.

Ointment jars

Ointment jars come in a variety of different sizes and can be made of either colourless glass or amber glass (Figure 1.6). Amber ointment jars are used for preparations that are sensitive to light. They are used to package ointments and creams and can be used for individually wrapped suppositories. As with cartons, additional care must be exercised in the storage of preparations in ointment jars as they do not come with child-resistant closures.

Collapsible tubes

Collapsible tubes come in a variety of different sizes and can be used to package creams or ointments. They are less convenient to fill than ointment jars and as such are rarely used for individual patient formulations. They are more commonly utilised in small-scale manufacturing environments where a number of identical products are being manufactured at the same time. As with ointment jars, additional care must be exercised in the storage of collapsible tubes as they do not come with child-resistant closures.

Standards for extemporaneous dispensing

Patients who visit a pharmacy with a prescription for a product needing to be extemporaneously prepared are entitled to expect the standards within a pharmacy to be comparable to those of a licensed manufacturing unit. The products produced within the pharmacy must be suitable for use, accurately prepared and prepared in such a way as to ensure the products meet the required standard for quality assurance. So, although this is small-scale

Figure 1.6.
A selection of ointment jars.

production, the same careful attention to detail is required as would be found in a manufacturing unit.

The following measures must be taken into consideration when preparing a product extemporaneously:

- **Personal hygiene.** Personal hygiene is extremely important. Hygiene standards within a pharmaceutical environment should be as high as, if not higher than, those found in food kitchens. This is because within a pharmaceutical environment, medication is being prepared for patients who may already be ill.
- **Personal protective equipment.** A clean white coat should be worn to protect the compounder from the product and conversely the product from contamination from the compounder. During the compounding process, safety glasses should always be worn and, depending on the nature of the ingredients to be incorporated into the preparation, additional safety equipment (e.g. face masks, gloves) may also be required. It is the responsibility of the individual compounder to assess the risk posed by any pharmaceutical ingredient and to ensure that the correct safety equipment is in use. Similarly, long hair should be tied back and hands washed, ensuring any open cuts are covered.
- **Clean work area and equipment.** The cleanliness of the work area and equipment used during the compounding procedure is of paramount importance. The risk of contaminating the final product with either dirt or microorganisms from the surroundings or from ingredients from a previous preparation can be considerable if attention is not paid to the cleanliness of the work area and equipment. Before starting to compound a product, the work area and equipment should be cleaned with a suitable solution (e.g. industrial denatured alcohol (IDA),

formerly known as industrial methylated spirits (IMS)), which must be allowed to dry fully.

- **Work area**. In addition to the cleanliness of the work area, consideration needs to be given to the work area itself to ensure that it is suitable for its intended purpose. Both lighting and ventilation need to be adequate. Some pharmaceutical ingredients are highly volatile and so, if the ventilation within the work area is inadequate, this could cause problems for the compounding staff.
- **Label preparation**. The label for any pharmaceutical product must be prepared before starting the compounding procedure. This will enable the product to be labelled as soon as it has been manufactured and packaged. This will eliminate a situation where an unlabelled product is left on the bench and would reduce the possibility of the product being mislabelled and given to the wrong patient.
- **Weighing and measuring procedure**. During weighing and measuring, unless strict guidelines are followed, it can be very easy to mix up different pharmaceutical ingredients as many ingredients resemble each other. It is preferable to incorporate a weighed or measured ingredient into a product as soon as possible to prevent any accidental switching. If this is not possible, when weighing or measuring more than one ingredient, place each on a piece of labelled paper as soon as it has been weighed or measured. This will avoid any accidental cross-over of ingredients.

chapter 2
Solutions

Overview

Upon completion of this chapter, you should be able to:

- prepare a solution from first principles
- select an appropriate container in which to package a solution
- prepare an appropriate label for a solution
- understand the differing calculations used in the preparation of solutions, including:
  Basic strength calculations
  Tailored strength calculations
  Percentage calculations
  Parts calculations
  Millimolar calculations.

Introduction and overview of solutions

Solutions are traditionally one of the oldest dosage forms used in the treatment of patients and afford rapid and high absorption of soluble medicinal products. Therefore, the compounding of solutions retains an important place in therapeutics today. Owing to the simplicity and therefore speed of preparation of an ad hoc formulation, they are of particular use for individuals who have difficulty in swallowing solid dosage forms (for example, paediatric, geriatric, intensive care and psychiatric patients), where compliance needs to be checked on administration (for example, in prisons or psychiatric pharmacy) and in cases where precise, individualised dosages are required.

Generally, water is chosen as the vehicle in which medicaments are dissolved, since it is non-toxic, non-irritant, tasteless, relatively cheap and many drugs are water-soluble. Problems may be encountered where active drugs are not particularly water-soluble or suffer from hydrolysis in aqueous solution. In these cases it is often possible to formulate a vehicle containing water mixed with a variety of other solvents.

Definition

Essentially a solution is a homogeneous liquid preparation that contains one or more dissolved medicaments. Since, by definition, active ingredients are dissolved within the vehicle, uniform doses by volume may be obtained without any need to shake the formulation. This is an advantage over some other formulation types, e.g. suspensions (see Chapter 3).

British Pharmacopoeia (BP) definition (oral solutions)

Oral solutions are oral liquids containing one or more active ingredients dissolved in a suitable vehicle.

KeyPoints

Advantages and disadvantages of solutions as dosage forms

Advantages
- Drug available immediately for absorption
- Flexible dosing
- May be designed for any route of administration
- No need to shake container
- Facilitates swallowing in difficult cases

Disadvantages
- Drug stability often reduced in solution
- Difficult to mask unpleasant tastes
- Bulky, difficult to transport and prone to container breakages
- Technical accuracy needed to measure dose on administration
- Some drugs poorly soluble
- Measuring device needed for administration

KeyPoints

Dissolution will normally take place in a glass beaker, not a conical measure, for a number of reasons.
- Firstly, owing to the shape of the conical measure, any solid added to a conical measure will tend to cake at the bottom of the measure and hamper any attempt to stir the solid around with the stirring rod which aids dissolution.
- Secondly, the action of the stirring rod may scratch the inside of the glass conical measure, permanently altering the internal volume of the measure.

General method

The following general method should be used in the preparation of a solution:

1. Write out the formula either from the prescription (unofficial) or from an official text (official).
2. Calculate the quantities required for each ingredient in the formula to produce the required final volume. Remember, it is not usual to calculate for an overage of product in the case of solutions as it is relatively easy to transfer the entire final contents of the conical measure. Additionally, as far as is practically possible, the product will be assembled in the final measure, thus reducing any transference losses.
3. Complete all sections of the product worksheet.
4. Prepare a suitable label.
5. Weigh all solids.
6. Identify the soluble solids and calculate the quantity of vehicle required to dissolve the solids fully. If more than one solid is to be dissolved, they are dissolved one by one, in order of solubility (i.e. the least soluble first). In almost all cases, dissolution will take place in a glass (or occasionally plastic) beaker, not a conical measure. Remember that the solubility of the soluble solids will be dependent on the vehicle used.
7. Transfer the appropriate amount of vehicle to a glass beaker.
8. If necessary, transfer the solid to a glass mortar and use the glass pestle to reduce particle size to aid dissolution (Figure 2.1).
9. Transfer the solid to the beaker and stir to aid dissolution. If a mortar and pestle have been used to reduce particle size, ensure that the mortar is rinsed with a little vehicle to ensure complete transfer of the powders.
10. When all the solid(s) has/have dissolved, transfer the solution to the conical measure that will be used to hold the final solution.

Figure 2.1. A porcelain mortar and pestle and a smaller glass mortar and pestle.

11. Rinse out the beaker in which the solution was made with a portion of the vehicle and transfer the rinsings to the conical measure.
12. Add any remaining liquid ingredients to the conical measure and stir.
13. Make up to final volume with remaining vehicle.
14. Transfer to a suitable container, label and dispense to the patient.

See Solutions video for a demonstration of the preparation of a solution.

Worked examples

Example 2.1
The preparation of Alkaline Gentian Mixture BP

You receive a prescription in your pharmacy with the following details:

Patient:	Mr David Shaw, 5 Longmeadow, Astonbury
Age:	56
Prescription:	Mist Gent Alk
Directions:	10 mL tds ac ex aq
Mitte:	150 mL

KeyPoints

During the dissolution phase, solutions should be stirred gently and uniformly to avoid air entrapment which may result in foaming of the solution. If available, automatic stirring devices may be useful in assisting the production of a uniform product and can be time-saving. If stirring devices are used to assist dissolution (e.g. rod, magnetic stirrers), remember to remove them before adjusting to final volumes.

Tips

It is best to stir continuously when combining ingredients into a solution (either liquid or solid ingredients). By stirring continually during incorporation, high concentration planes within the fluid body, which might increase the likelihood of incompatibilities, will be avoided.

Tips

Further considerations during the preparation of a solution:
1. To aid dissolution, high-viscosity liquid components should be added to those of lower viscosity.
2. Completely dissolve salts in a small amount of water prior to the addition of other solvent elements.
3. In complex solutions, organic components should be dissolved in alcoholic solvents and water-soluble components dissolved in aqueous solvents.
4. Aqueous solutions should be added to alcoholic solutions with stirring to maintain the alcohol concentration as high as possible – the reverse may result in separation of any dissolved components.

1. **Use of the product**
    Gentian is used as a bitter to stimulate appetite (*British Pharmacopoeia* 2007, p 949).
2. **Is it safe and suitable for the intended purpose?**
    This is an official preparation, therefore the formula is safe and suitable for purpose. Note that if the dose of an oral liquid is specified as 5 mL or 10 mL, the dose regimen would be 5 mL or 10 mL three or four times daily by convention.
3. **Calculation of formula for preparation**
    Prepare 150 mL of Alkaline Gentian Mixture BP.

Product formula
(from the British Pharmacopoeia 2007, p 2616):

	Master	100 mL	50 mL	150 mL
Concentrated Compound Gentian Infusion BP	100 mL	10 mL	5 mL	15 mL
Sodium Bicarbonate BP	50 g	5 g	2.5 g	7.5 g
Double Strength Chloroform Water BP	500 mL	50 mL	25 mL	75 mL
Potable water	to 1000 mL	to 100 mL	to 50 mL	to 150 mL

Interim formula for Double Strewngth Chloroform Water BP

Concentrated Chloroform Water BPC 1959	5 mL
Potable water	to 100 mL

4. **Method of preparation**
a. Solubility where applicable
    Sodium Bicarbonate BP is soluble 1 in 11 in water (*British Pharmacopoeia* 1988, p 509). Therefore to dissolve 7.5 g Sodium Bicarbonate BP, a minimum of 7.5 × 11 = 82.5 mL of water would be required. As this is greater than 50% of the mixture, a solution of Double Strength Chloroform Water BP and water would be used for dissolution.
b. Vehicle/diluent
    Double Strength Chloroform Water BP and potable water would be used as the vehicle as per the product formula.
c. Preservative
    Double Strength Chloroform Water BP is included in this product as the preservative as per the product formula.
d. Flavouring when appropriate
    No extra flavouring is required. In addition to preservative action Double Strength Chloroform Water BP will give some flavouring and act as a sweetener, therefore helping to counteract the bitter taste of the Concentrated Compound Gentian Infusion BP

The following method would be used to prepare 150 mL of Alkaline Gentian Mixture BP from the formula above:

1. Using the master formula from the *British Pharmacopoeia* for 1000 mL of final product, calculate the quantity of ingredients required to produce the final volume needed (150 mL).
2. Calculate the composition of a convenient quantity of Double Strength Chloroform Water BP, sufficient to satisfy the formula requirements but also enabling simple, accurate measurement of the concentrated component.

Method of compounding for Double Strength Chloroform Water BP

a. In this case, 75 mL of Double Strength Chloroform Water BP is required and so it would be sensible to prepare 100 mL. To prepare 100 mL Double Strength Chloroform Water BP, measure 5 mL of Concentrated Chloroform water BPC 1959 accurately using a 5 mL conical measure.
b. Add approximately 90 mL of potable water to a 100 mL conical measure (i.e. sufficient water to enable dissolution of the concentrated chloroform component without reaching the final volume of the product).
c. Add the measured Concentrated Chloroform Water BPC 1959 to the water in the conical measure.
d. Stir gently and then accurately make up to volume with potable water.
e. Visually check that no undissolved chloroform remains at the bottom of the measure.

Noting that Sodium Bicarbonate BP is soluble 1 in 11 with water, a minimum of 11 mL of water would be required to dissolve 1 g of Sodium Bicarbonate BP.

The final volume of Alkaline Gentian Mixture BP required (150 mL) will contain 7.5 g of Sodium Bicarbonate BP. As 1 g of Sodium Bicarbonate BP is soluble in 11 mL, 7.5 g is soluble in 82.5 mL (7.5 × 11 = 82.5 mL).

Therefore a minimum of 82.5 mL of vehicle would be required to dissolve the 7.5 g of Sodium Bicarbonate BP in this example. For ease of compounding, choose a convenient volume of vehicle, say 90 mL, in which to dissolve the solute initially. When choosing the amount of vehicle to use for

Tips

Rather than attempt the conversion from 1000 mL to 150 mL in one stage, it may be simpler to take the calculation through a number of stages. In the example given above, the quantities in the master formula are first divided by 10 to give a product with a final volume of 100 mL. These quantities are then halved to give a product with a final volume of 50 mL. The quantities in the 100 mL product and 50 mL product are then added together to give the quantities of ingredients in a product with a final volume of 150 mL. By using this method, the compounder is less likely to make a calculation error.

Tips

As discussed above, in this example 90 mL of vehicle is required to dissolve the Sodium Bicarbonate BP. It is important to consider the total amount of each liquid ingredient in the product to ensure that only the correct amounts are added.

In this example, it would be incorrect to dissolve the Sodium Bicarbonate BP in 90 mL of Double Strength Chloroform Water BP as the final volume of the preparation only contains 75 mL. Equally, it would also be incorrect to dissolve the Sodium Bicarbonate BP in 90 mL of water as the final volume of the preparation will contain less than 75 mL.

In this case, all the Double Strength Chloroform Water BP is used (75 mL) along with enough potable water to reach the desired volume (approximately 15 mL).

dissolution, it is important to consider the total amount of each liquid ingredient in the preparation to ensure that only the correct amounts are added or the final product does not go over volume.

3. Weigh 7.5 g Sodium Bicarbonate BP on a Class II (Figure 2.2) or electronic balance.
4. Accurately measure 75 mL Double Strength Chloroform Water BP using a 100 mL measure. To this add approximately 15 mL potable water in order to produce 90 mL of vehicle which should be poured into a beaker (in order to produce sufficient volume to dissolve the 7.5 g Sodium Bicarbonate BP).
5. The Sodium Bicarbonate BP (7.5 g) should be added to the vehicle, thus following the principle of adding solutes to solvents.
6. Stir to aid dissolution.
7. Transfer the solution to a 250 mL conical measure.
8. Rinse the beaker with potable water, adding the rinsings to the Sodium Bicarbonate BP solution.
9. Accurately measure 15 mL of Concentrated Compound Gentian Infusion BP in an appropriately sized conical measure and add to the Sodium Bicarbonate BP solution in the 250 mL measure. Rinse out

Figure 2.2.
A Class II balance.

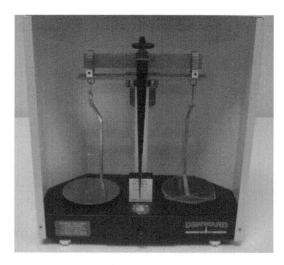

the small conical measure with potable water and add the rinsings to the mixture.

10. Make up to volume (150 mL) accurately with potable water and stir.

11. Transfer the solution to a 150 mL amber flat medical bottle with a child-resistant closure and label.

5. **Choice of container**
   A plain amber bottle with a child-resistant closure would be most suitable as the preparation is a solution for internal use.

6. **Labelling considerations**

a. Title
   The product is official, therefore the following title would be suitable: 'Alkaline Gentian Mixture BP'.

b. Quantitative particulars
   Quantitative particulars are not required as the product is official.

c. Product-specific cautions (or additional labelling requirements)
   Not applicable.

d. Directions to patient – interpretation of Latin abbreviations where necessary
   'Take TWO 5 mL spoonfuls THREE times a day before food in water'.

e. Recommended *British National Formulary* cautions when suitable
   Not applicable.

f. Discard date
   The British Pharmacopoeia (2004, p 2453) states that this product should be recently prepared, therefore it will attract a 4-week discard date. Alternatively, a 2-week discard date could be assigned as the product also contains an infusion.

g. Sample label (you can assume that the name and address of the pharmacy and the words 'Keep out of the reach and sight of children' are pre-printed on the label):

Alkaline Gentian Mixture BP	150 mL
Take TWO 5 mL spoonfuls THREE times a day before food in water.	
Do not use after (2 weeks)	
Mr David Shaw	Date of dispensing

7. **Advice to patient**
   The patient would be advised to mix two 5 mL spoonfuls with an equal volume of water to dilute the bitter taste three times a day before food. In addition, the discard date would be highlighted to the patient.

Example 2.2
The preparation of Ammonium Chloride Mixture BP

You receive a prescription in your pharmacy with the following details:

Patient:	Mr James Watson, 4 Arrow Ave, Astonbury
Age:	42
Prescription:	Ammonium Chloride Mixture BP
Directions:	10 mL tds prn
Mitte:	50 mL

1. **Use of the product**
   Used as an expectorant to treat chesty coughs (*Martindale* 35th edn, p 1399).
2. **Is it safe and suitable for the intended purpose?**
   This is an official preparation, therefore the formula is safe and suitable for purpose. The dose of 10 mL three times a day when required is consistent with the adult standard dosage for official oral liquids (of 10 mL) three to four times a day by convention.
3. **Calculation of formula for preparation**
   Prepare 50 mL of Ammonium Chloride Mixture BP.

Product formula
(from the *British Pharmacopoeia* 2007, p 2317)

	Master	500 mL	50 mL
Ammonium Chloride BP	100 g	50 g	5 g
Aromatic Ammonia Solution BP	50 mL	25 mL	2.5 mL
Liquorice Liquid Extract BP	100 mL	50 mL	5 mL
Potable water	to 1000 mL	to 500 mL	to 50 mL

4. **Method of preparation**
a. Solubility where applicable
   Ammonium Chloride BP is soluble 1 in 2.7 in water (*British Pharmacopoeia* 1988, p 36). Therefore to dissolve 5 g of ammonium chloride a minimum of 5 × 2.7 = 13.5 mL of water would be required.
b. Vehicle/diluent
   Potable water would be used as the vehicle as per the product formula.
c. Preservative
   No additional preservative is required as per the product formula.
d. Flavouring when appropriate

Liquorice Liquid Extract BP, although a mild expectorant, is mainly included for its flavouring and sweetening properties and its ability to disguise the taste of the Ammonium Chloride BP.

Method of compounding

1. Calculate the quantity of ingredients required to produce the final volume needed. As with Example 2.1, this calculation is best attempted in stages.
2. Weigh 5 g Ammonium Chloride BP accurately on a Class II or electronic balance.
3. Measure approximately 15 mL potable water and transfer to a beaker.
4. Add the Ammonium Chloride BP to the water in the beaker and stir until dissolved.
5. Transfer to a 50 mL conical measure with rinsings.
6. Measure 5 mL Liquorice Liquid Extract BP accurately in a 5 mL conical measure and add, with rinsings, to the 50 mL measure containing ammonium chloride solution.
7. Measure 2.5 mL Aromatic Ammonia Solution BP accurately in a syringe and transfer to the 50 mL measure containing the composite solution.
8. Make up to the final volume of 50 mL with potable water and stir.
9. Pack into a 50 mL amber flat medicine bottle and label.

Tips

As Ammonium Chloride BP is soluble 1 part in 2.7 parts of water, the 5 g required for this product would only dissolve in a minimum initial volume of 13.5 mL aqueous vehicle. Therefore we should choose a convenient volume of vehicle to dissolve the solute, for example, 15 mL.

Tips

The solution should be gently stirred to reduce the likelihood of frothing that can occur upon the addition of Liquorice Liquid Extract BP. Any resultant frothing will make the accurate reading of the final meniscus difficult.

Tips

Note that the syringe will not need rinsing as it is designed to deliver a measured volume.

5. **Choice of container**

   A plain amber bottle with a child-resistant closure would be the most suitable as the preparation is a solution intended for internal use.

6. **Labelling considerations**

   a. Title

   The product is official, therefore the following title would be suitable: 'Ammonium Chloride Mixture BP'.

   b. Quantitative particulars

   Quantitative particulars are not required as the product is official.

   c. Product-specific cautions (or additional labelling requirements)

   No product-specific cautions are applicable, although the patient could be advised to take the product in water (to dilute the taste) or in warm water (to dilute the taste but also to improve the expectorant action by hastening the release of the ammonia, which is the main expectorant in the mixture).

d. Directions to patient – interpretation of Latin abbreviations where necessary
'Take TWO 5 mL spoonfuls THREE times a day when required'.

e. Recommended *British National Formulary* cautions when suitable
Not applicable.

f. Discard date
The *British Pharmacopoeia* (2004, p 2181) states that this product should be recently prepared, therefore it will attract a 4-week discard date.

g. Sample label (you can assume that the name and address of the pharmacy and the words 'Keep out of the reach and sight of children' are pre-printed on the label):

Ammonium Chloride Mixture BP	**50 mL**
Take TWO 5 mL spoonfuls THREE times a day when required.	
Do not use after (4 weeks)	
Mr James Watson	Date of dispensing

7. **Advice to patient**
The patient would be advised to take two 5 mL spoonfuls three times a day when required. In addition, the discard date and the fact that the solution may be taken diluted with an equal volume of warm water would be highlighted.

Example 2.3
The preparation of Sodium Chloride Compound Mouthwash BP

You receive a prescription in your pharmacy with the following details:

Patient:	Mrs Avril Asker, 21 Station Road, Astonbury
Age:	36
Prescription:	Sodium Chloride Compound Mouthwash BP
Directions:	Use 20 mL qqh prn
Mitte:	150 mL

1. **Use of the product**
Used as a mouthwash to cleanse and freshen the mouth, suitable for treatment of superficial infections if used frequently (*British National Formulary* 61st edn, p 697).

2. **Is it safe and suitable for the intended purpose?**
   This is an official preparation, therefore the formula is safe and suitable for purpose. The dose of 20 mL every 4 hours when required is consistent with the recommended dosage (*British National Formulary* 61st edn, p 698).
3. **Calculation of formula for preparation**
   Prepare 150 mL of Sodium Chloride Compound Mouth wash BP.

Product formula
(from the *British Pharmacopoeia* 2007, p 2911)

	Master	100 mL	50 mL	150 mL
Sodium Bicarbonate BP	10 g	1 g	500 mg	1.5 g
Sodium Chloride BP	15 g	1.5 g	750 mg	2.25 g
Concentrated Peppermint Emulsion BP	25 mL	2.5 mL	1.25 mL	3.75 mL
Double Strength Chloroform Water BP	500 mL	50 mL	25 mL	75 mL
Potable water	to 1000 mL	to 100 mL	to 50 mL	to 150 mL

Interim formula for Double Strength Chloroform Water BP

Concentrated Chloroform Water BPC 1959	5 mL
Potable water	to 100 mL

4. **Method of preparation**
a. Solubility where applicable
   Sodium Bicarbonate BP is soluble 1 in 11 parts of water (*British Pharmacopoeia* 1988, p 509). Therefore to dissolve 1.5 g of Sodium Bicarbonate BP a minimum of $1.5 \times 11 = 16.5$ mL water would be required. Sodium Chloride BP is soluble 1 in 3 parts of water (*British Pharmacopoeia* 1988, p 512). Therefore to dissolve 2.25 g of Sodium Chloride BP a minimum of $2.25 \times 3 = 6.75$ mL water would be required.
b. Vehicle/diluent
   Double Strength Chloroform Water BP and potable water would be used as the vehicle as per the product formula.
c. Preservative
   Double Strength Chloroform Water BP is included in this product as the preservative as per the product formula.
d. Flavouring when appropriate
   Concentrated Peppermint Emulsion BP is added to this product as a flavouring as per the product formula. In addition the Double Strength Chloroform Water BP will also sweeten and flavour the product. The following method would be used

to prepare 150 mL of Sodium Chloride Compound Mouthwash BP from the formula above:

1. Using the master formula from the *British Pharmacopoeia* for 1000 mL of final product, calculate the quantity of ingredients required to produce the final volume needed (150 mL).
2. Calculate the composition of a convenient quantity of Double Strength Chloroform Water BP, sufficient to satisfy the formula requirements but also enabling simple, accurate measurement of the concentrated component.

Method of compounding for Double Strength Chloroform Water BP

a. In this case, 75 mL of Double Strength Chloroform Water BP is required and so it would be sensible to prepare 100 mL. To prepare 100 mL Double Strength Chloroform Water BP, measure 5 mL of Concentrated Chloroform Water BPC 1959 accurately using a 5 mL conical measure.
b. Add approximately 90 mL of potable water to a 100 mL conical measure (i.e. sufficient water to enable dissolution of the concentrated chloroform component without reaching the final volume of the product).
c. Add the measured Concentrated Chloroform Water BPC 1959 to the water in the conical measure.
d. Stir gently and then accurately make up to volume with potable water.
e. Visually check that no undissolved chloroform remains at the bottom of the measure.

Note that Sodium Bicarbonate BP is soluble 1 in 11 with water, a minimum of 11 mL of water would be required to dissolve 1 g of Sodium Bicarbonate BP.

The final volume of Sodium Chloride Compound Mouthwash BP required (150 mL) will contain 1.5 g of Sodium Bicarbonate BP. As 1 g of Sodium Bicarbonate BP is soluble in 11 mL, 1.5 g is soluble in 16.5 mL (1.5 × 11 = 16.5 mL).

The Sodium Chloride BP is soluble 1 in 2.8 with water. Therefore a minimum of 2.8 mL of water would be required to dissolve 1 g of Sodium Chloride BP.

The final volume of Sodium Chloride Compound Mouthwash BP required (150 mL) will contain 2.25 g of Sodium Chloride BP. As 1 g of Sodium Chloride BP is soluble in 2.8 mL, 2.25 g is soluble in 6.3 mL (2.25 × 2.8 = 6.3 mL).

Therefore a minimum of 16.5 mL of vehicle would be required to dissolve the 1.5 g of Sodium Bicarbonate BP and a minimum of 6.3 mL of vehicle would be required to dissolve the 2.25 g of Sodium Chloride BP in this example.

For ease of compounding, choose a convenient volume of vehicle, for example, 30 mL, in which to dissolve the solute initially.

3. Weigh 2.25 g Sodium Chloride BP on a Class II or electronic balance.
4. Weigh 1.5 g Sodium Bicarbonate BP on a Class II or electronic balance.
5. Measure 75 mL of Double Strength Chloroform Water BP in a 100 mL conical measure.
6. Transfer approximately 30 mL of Double Strength Chloroform Water BP to a beaker; add the Sodium Bicarbonate BP and stir to aid dissolution.
7. When the Sodium Bicarbonate BP has dissolved, add the Sodium Chloride BP and stir to aid dissolution.
8. Transfer the solution from the beaker to a 250 mL conical measure.
9. Rinse out the beaker with some Double Strength Chloroform Water BP and add the rinsings to the conical measure.
10. Measure 3.75 mL of Concentrated Peppermint Emulsion BP using a 5 mL and a 1 mL syringe.
11. Add the Concentrated Peppermint Emulsion BP to the conical measure.
12. Make up to volume with the remaining Double Strength Chloroform Water BP and potable water.
13. Transfer to a 200 mL amber fluted bottle and label.

Tips

When choosing the amount of vehicle to use for dissolution, it is important to consider the total amount of each liquid ingredient in the preparation to ensure that only the correct amounts are added or the final product does not go over volume.

5. **Choice of container**
    As this is an extemporaneously prepared mouthwash, the container of choice would be an amber fluted medical bottle with child-resistant closure.
6. **Labelling considerations**
a. Title
    The product is official, therefore the following title would be suitable: 'Sodium Chloride Compound Mouthwash BP'.
b. Quantitative particulars
    Quantitative particulars are not required as the product is official.
c. Product-specific cautions
    As the product is a mouthwash, 'Not to be taken', 'Do not swallow in large amounts' and 'Dilute with an equal volume of water before using' would all be required on the product label.
d. Directions to patient – interpretation of Latin abbreviations where necessary
    'Use 20 mL as a mouthwash every four hours when required'.

e. Recommended *British National Formulary* cautions when suitable
The *British National Formulary* has no specific labelling cautions for this product but does advise dilution with an equal volume of warm water.

f. Discard date
The product is a mouthwash containing a preservative and so will attract a 4-week discard date.

g. Sample label (you can assume that the name and address of the pharmacy and the words 'Keep out of the reach and sight of children' are pre-printed on the label):

Sodium Chloride Compound Mouthwash BP	**150 mL**
Use 20 mL as a mouthwash every four hours when required.	
Dilute with an equal volume of water before use.	
Do not swallow in large amounts	
Not to be taken	
Do not use after (4 weeks)	
Mrs Avril Asker	Date of dispensing

7. **Advice to patient**
The patient would be advised that 20 mL of mouthwash should preferably be diluted with an equal volume of warm water and used every 4 hours when required. In addition, the discard date and the fact that, although the mouthwash should not be swallowed, it is not harmful to swallow small amounts would be highlighted.

Example 2.4
The preparation of a magistral formulation from a doctor's prescription

You receive a prescription in your pharmacy with the following details:

Patient:	Mr Gary Murray, 12 Bishop Road, Astonbury
Age:	49
Prescription:	Potassium Permanganate Solution 0.2%
Directions:	Dilute 1 in 20 and use as a wet dressing alt die
Mitte:	150 mL

1. **Use of the product**
Used for cleansing wounds and deodorising suppurating eczematous reactions and wounds (*British National Formulary* 61st edn, p 744).

2. **Is it safe and suitable for the intended purpose?**
   It is commonly used as a solution at a strength of 1 in 10
   000 (*British National Formulary* 61st edn, p 744). When the
   prepared solution is diluted as indicated it will provide a 1 in
   10 000 solution.
       0.2% w/v solution is the same as a 1 in 500 solution
   which, when diluted 20 times, becomes a 1 in 10 000 solution.
3. **Calculation of formula for preparation**
   Prepare 150 mL of Potassium Permanganate Solution 0.2% w/v.
       0.2% w/v is equal to 0.2 g in 100 mL. Therefore there is
   200 mg of Potassium Permanganate BP in every 100 mL of
   solution.

Product formula

	Master	50 mL	150 mL
Potassium Permanganate BP	200 mg	100 mg	300 mg
Freshly boiled and cooled purified water	to 100 mL	to 50 mL	to 150 mL

Tips

4. **Method of preparation**
a. Solubility where applicable
   Potassium Permanganate BP is soluble 1
   in 16 in cold water and freely soluble in
   boiling water (*British Pharmacopoeia* 1988,
   p 455).
b. Vehicle/diluent
   Freshly boiled and cooled purified water
   would be used as the vehicle, as no
   preservative will be added and the solution
   is intended for application to a wound.
c. Preservative
   No preservative is to be added to this
   product.
d. Flavouring when appropriate
   Not applicable as the solution is for
   external use.

Rather than attempt the above
conversion in one stage, it may
be simpler to take the calculation
through a number of stages. In
the example given above, the
quantities in the master formula
are first divided by 2 to give a
product with a final volume of
50 mL. The quantities in the
100 mL product and 50 mL product
are then added together to give
the quantities of ingredients in
a product with a final volume of
150 mL. Using this method, the
compounder is less likely to make
a calculation error.

The following method would be used to prepare 150 mL of
Potassium Permanganate BP 0.2% w/v from the formula above:
1. Weigh 300 mg Potassium Permanganate BP on a Class II or
   electronic balance.
2. Transfer to a glass mortar as the Potassium Permanganate BP
   is crystalline and for ease of dissolution needs to be ground
   under water into a powder.
3. Transfer the solution to a 250 mL conical measure.

Tips

Potassium Permanganate BP is an oxidising substance, therefore there is risk of explosion. To prevent this, add approximately 20 mL of freshly boiled and cooled purified water to a glass mortar (Potassium Permanganate BP stains and so a porcelain mortar would not be suitable) and grind under water. In addition, a glass mortar is always used for grinding as crystalline substances can scratch a porcelain mortar.

4. Rinse the mortar with freshly boiled and cooled purified water and add the rinsings to the conical measure.
5. Make up to volume with freshly boiled and cooled purified water.
6. Transfer the solution to a 150 mL amber fluted medical bottle with a child-resistant closure and label.

5. **Choice of container**

    A fluted amber bottle with a child-resistant closure would be most suitable as the preparation is a solution for external use.

6. **Labelling considerations**

    a. Title

        The product is unofficial, therefore the following title would be suitable: 'Potassium Permanganate Solution 0.2% w/v'.

    b. Quantitative particulars

        The product is unofficial, therefore it is necessary to put the quantitative particulars on the label. As the product is intended for external use, the quantitative particulars would be expressed per container.

This container contains:

Potassium Permanganate BP	0.2%
Freshly boiled and cooled purified water	to 100%

or

This container contains:

Potassium Permanganate BP	300 mg
Freshly boiled and cooled purified water	to 150 mL

The quantitative particulars for a product for external use may be expressed in percentage terms instead of actual quantities; either would be correct.

    c. Product-specific cautions (or additional labelling requirements).

        'For external use only.'

    d. Directions to patient – interpretation of Latin abbreviations where necessary

        'Dilute ONE capful with NINETEEN capfuls of water and use as a wet dressing every other day.'

    e. Recommended *British National Formulary* cautions when suitable

The *British National Formulary* has no specific labelling cautions for this product but does advise that solutions of Potassium Permanganate BP stain skin and clothing. Therefore, a suitable caution could be added to the final label.

f. Discard date

The product would attract a 2-week discard date as there is no preservative and the product is being applied to an open wound.

g. Sample label (you can assume that the name and address of the pharmacy and the words 'Keep out of the reach and sight of children' are pre-printed on the label):

Potassium Permanganate Solution 0.2%w/v   **150 mL**

Dilute ONE capful with NINETEEN capfuls of water and use as
a wet dressing every other day.
Caution: Stains skin, hair and fabric
Do not use after (2 weeks)
For external use only

This solution contains:

Potassium Permanganate BP	0.2%
Freshly boiled and cooled purified water	to 100%

Mr Gary Murray   Date of dispensing

---

7. **Advice to patient**

The patient would be advised to dilute one capful of the solution with 19 capfuls of water and use as a wet dressing every other day. In addition, the discard date and the following cautions would be highlighted to the patient:

For external use only.

The preparation may stain hair, skin and fabrics.

With prolonged use nails may also be stained.

Discontinue use if the skin becomes dry.

Self-assessment

Basic strength calculations

The simplest way to express the strength of a solution is to specify the amount of solute to be dissolved in a stated amount of solvent.

If the solute is a solid dissolved in a liquid, the strength of the solution may often be expressed as mg/mL, mg/100 mL, g/100 mL, mg/L or g/L. Similarly, if the solute is a liquid, the strength could be expressed as mL/10 mL, mL/100 mL or mL/L.

Example 2.5

You are asked to prepare a 100 mL solution containing Sodium Chloride BP 9 mg/mL

Sodium Chloride BP	9 mg	90 mg	900 mg
Potable water	to 1 mL	to 10 mL	to 100 mL

Therefore the amount required would be 900 mg (= 0.9 g).

Similarly, the request could be to prepare 100 mL of a solution containing Sodium Chloride BP 0.009 g/mL

Sodium Chloride BP	0.009 mg	0.09 mg	0.9 mg
Potable water	to 1 mL	to 10 mL	to 100 mL

Therefore the amount required would be 0.9 g (= 900 mg).

Questions

1. How much solid would be required in order to produce 500 mL of a 15 mg/10 mL solution?
   a. 75 mg
   b. 150 mg
   c. 750 mg
   d. 1500 mg
   e. 7500 mg

2. If 30 mg of an ingredient was dissolved in 1.5 mL of solvent, what would be the strength of the resulting solution expressed as mg/mL?
   a. mg/mL
   b. 15 mg/mL
   c. 20 mg/mL
   d. 30 mg/mL
   e. 200 mg/mL

3. A patient requires a dose of 1 mg of atropine sulphate. Ampoules containing 600 micrograms/mL are available. If a 2 mL syringe graduated to 0.1 mL is available, which of the following provides the nearest dose?
   a. 1.5 mL
   b. 1.6 mL
   c. 1.7 mL
   d. 1.8 mL
   e. 1.9 mL

4. A paediatric vitamin drop contains 0.25 mg of vitamin D in each millilitre. How many micrograms of vitamin D are contained in 0.2 mL of this preparation?
   a. 50 micrograms
   b. 75 micrograms
   c. 100 micrograms
   d. 150 micrograms
   e. 250 micrograms

5. **What weight of sodium bicarbonate (in grams) would be required to make 150 mL of a 6 g/L solution?**
   a. 0.5 g
   b. 0.6 g
   c. 0.75 g
   d. 0.9 g
   e. 1 g

Tailored strength calculations

Often this type of calculation is required if you are attempting to give a tailored dose to a patient using existing pre-prepared stock mixtures.

Example 2.6

A common dose seen in paediatric prescribing is 62.5 mg phenoxymethylpenicillin four times a day. This is the recommended dose for a child 1 month–1 year. The readily available mixture is 125 mg/5 mL. Therefore to provide a dose of 62.5 mg we give 2.5 mL of a 125 mg/mL mixture.

$$\text{Volume required} = \frac{\text{strength required}}{\text{stock strength}} \times \text{volume of stock solution}$$

$$= \frac{62.5}{125} \times 5 \text{ ml}$$

$$= \frac{62.5}{125} \times \frac{5 \text{ ml}}{1}$$

$$= \frac{312.5 \text{ ml}}{125}$$

$$= 2.5 \text{ ml}$$

Questions

6. A patient requires a dose of 5 mg of a drug. The available stock solution contains 25 mg/5 mL. How much of this stock solution would be required to deliver this dose?

7. A baby requires a dose of 37.5 mg chloroquine base each week to prevent infection with malarial parasite. The solution available for you to dispense contains 50 mg/5 mL chloroquine base. How much of this stock solution should be given to the baby each week?

8. A 5-year-old child needs a dose of 125 mg cimetidine four times a day. The stock solution of cimetidine available contains 200 mg/5 mL. How many millilitres of this stock solution will be administered for each dose?

Percentage calculations

Percentages are also commonly used to express the strength of solutions. Usually these solutions are not intended for the oral route of administration. As a percentage this can have four different meanings and in order to make clear the intention the following terms are used:

- % w/w percentage weight in weight. This expresses the amount in grams of solute in 100 g of product.
- % w/v percentage weight in volume. This expresses the amount in grams of solute in 100 mL of product.
- % v/v percentage volume in volume. This expresses the number of millilitres of solute in 100 mL of product.
- % v/w percentage volume in weight. This expresses the number of millilitres of solute in 100 g of product.

The strength of solutions of solids in liquids is usually expressed as % w/v, whereas that of liquids in liquids is expressed as % v/v. When the type of percentage is not specified by convention the above rule will apply. For example, % solid in liquid is interpreted as % w/v.

Example 2.7

Prepare 50 mL potassium permanganate solution 2.8%.

As Potassium Permanganate BP is a solid this would mean: prepare a solution containing potassium permanganate 2.8% w/v.

This means that there would be 2.8 g of Potassium Permanganate BP dissolved in every 100 mL of solution.

Potassium Permanganate BP	2.8 g	1.4 g
Freshly boiled and cooled purified water	to 100 mL	to 50 mL

Questions
What quantities would be required for the following?

9. 500 mL of a 0.1% w/v solution using a 20% w/v solution.

10. 5 L of a 0.9% w/v solution.

11. 20 mL of a 5% solution.

12. How much solid would be required in order to produce 50 mL of a 0.2% w/v solution?
    a. 100 micrograms
    b. 200 micrograms
    c. 10 milligrams
    d. 100 milligrams
    e. 200 milligrams

13. How much solid would be required in order to produce 20 mL of a 5% w/v solution?
    a. 100 mg
    b. 500 mg

c. 1000 mg
d. 5000 mg
e. 10 000 mg

14. **How much solid would be required in order to produce 300 mL of a 0.01% w/v solution?**
a. 3 micrograms
b. 3 milligrams
c. 30 milligrams
d. 300 milligrams
e. 3 grams

15. **How much solid would be required in order to produce 750 mL of a 15% w/v solution?**
a. 75 mg
b. 125 mg
c. 1125 mg
d. 112.5 g
e. 750 g

16. **How much solid would be required in order to produce 10 litres of a 0.45% solution?**
a. 45 mg
b. 4.5 g
c. 9 g
d. 45 g
e. 90 g

17. **What is the percentage strength when 5 mL of disinfectant concentrate is made up to 1 litre with water?**
a. 0.05% v/v
b. 0.15% v/v
c. 0.5% v/v
d. 1.5% v/v
e. 5% v/v

18. **You have been given the following prescription: Sodium Bicarbonate BP 5% potable water to 10 mL. How much Sodium Bicarbonate BP (in grams) will be needed to prepare the product?**
a. 0.05 g
b. 0.15 g
c. 0.5 g
d. 1.5 g
e. 5 g

19. **How much of a 20% w/v solution would be required to produce 250 mL of a 0.5% w/v solution?**
   a. 2.5 mL
   b. 6.25 mL
   c. 25 mL
   d. 62.5 mL
   e. 125 mL

20. **How much of a 4% w/v solution would be required to prepare 150 mL of a 1% w/v solution?**
   a. 2.6 mL
   b. 3.75 mL
   c. 26 mL
   d. 37 mL
   e. 37.5 mL

21. **How much of a 25% w/v solution would be required to prepare 250 mL of a 0.5% w/v solution?**
   a. 0.5 mL
   b. 1 mL
   c. 2.5 mL
   d. 5 mL
   e. 10 mL

22. **Calculate the amount of stock solution that would be required to make the following solutions.**
   a. Half a litre of a 1% v/v solution using a 15% v/v solution
   b. 250 mL of a 1% v/v solution using a 40% v/v solution
   c. 500 mL of a 1% v/v solution using a 10% solution
   d. 1 litre of a 0.5% v/v solution using a 15% solution
   e. 1 litre of a 0.05% solution using a 4% solution

Parts calculations

The concentration of solutions may also be expressed in terms of 'parts'. By this we mean 'parts' of solute in 'parts' of product. This is interpreted as parts by weight (grams) of a solid in parts by volume (millilitres) of the final solution or in parts by volume (millilitres) of a liquid in parts by volume (millilitres) of the final solution. Solubility of ingredients is often expressed in this way.

Example 2.8

Sodium Bicarbonate BP is soluble in 11 parts of water. This means that 1 g of Sodium Bicarbonate BP will dissolve in 11 mL of water. Therefore if you had a formula that required 4 g of Sodium Bicarbonate BP you would need a minimum of $4 \times 11 = 44$ mL of water in which to dissolve the 4 g of Sodium Bicarbonate BP.

Questions

23. How many millilitres of potable water are required to dissolve 3 g of a solid which is soluble in 2.5 parts of water?

a. 2.5 mL
b. 3 mL
c. 5 mL
d. 7.5 mL
e. 8 mL

24. How many millilitres of potable water are required to dissolve 7 g of a solid, which is soluble 1 in 1.5 parts of water?

a. 1.5 mL
b. 15 mL
c. 7.5 mL
d. 10.5 mL
e. 75 mL

25. Sodium Bicarbonate BP is soluble 1 in 11. How much Double Strength Chloroform Water BP is needed to dissolve 0.37 kg?

a. 3.7 L
b. 4.07 L
c. 4090 mL
d. 4.1 L
e. 4700 mL

26. How much of a 1 in 150 w/v solution would be required to produce 200 mL of a 0.2% solution?

a. 3 mL
b. 15 mL
c. 25 mL
d. 30 mL
e. 60 mL

27. Express 500 micrograms in 2 mL as '1 in x'.

a. 1 in 250
b. 1 in 400
c. 1 in 2000
d. 1 in 2500
e. 1 in 4000

28. How much Crystal Violet BP is required to prepare 5.4 litres of a 1 in 12 000 solution?

a. 450 micrograms
b. 540 micrograms
c. 4.5 milligrams
d. 450 milligrams
e. 540 milligrams

Millimolar calculations

The strength of active ingredient within a pharmaceutical preparation can be expressed as the number of millimoles per unit volume or mass of product. The mole is the unit of amount of substance and there are 1000 millimoles in a mole. To calculate the number of millimoles of an ingredient in a medicinal product, you will firstly need to know the molecular weight of the ingredient.

The number of moles of ingredient is the mass of ingredient divided by the molecular mass:

$$\text{Number of moles} = \frac{\text{Mass in grams}}{\text{Molecular mass}}$$

For example, the molecular weight quoted for Sodium Chloride BP is 58.44.

Therefore a molar solution of Sodium Chloride BP would contain 58.44 g of Sodium Chloride BP in a litre.

Example 2.9

Prepare 100 mL of Sodium Chloride BP solution containing 1.5 mmol per mL.

1 ml contains 1.5 mmol	
100 ml contains 150 mmol	
1 mole (1000 mmol) of Sodium Chloride BP weighs	58.44 g
1 mmol of Sodium Chloride BP weighs	$\dfrac{58.44 \text{ g}}{1000}$
150 mmol of Sodium Chloride BP weighs	= 8.766 g (weigh 8.77 g)

Questions

29. The molecular weight of Sodium Chloride BP is 58.44. How many grams of Sodium Chloride BP would be needed to prepare 1 litre of a molar solution?
   a. 0.2922 g
   b. 0.5844 g
   c. 5.844 g
   d. 29.22 g
   e. 58.44 g

30. The molecular weight of Sodium Bicarbonate BP is 84. How many grams of Sodium Bicarbonate BP would be required to produce 150 mL of 0.5 mmol/mL solution?
   a. 0.63 g
   b. 0.84 g
   c. 6.3 g
   d. 8.4 g
   e. 63 g

31. The molecular weight of Sodium Chloride BP is 58.44. How many grams of Sodium Chloride BP would be required to produce 100 mL of a 2 mmol/mL solution?

a. 2.92 g
b. 5.84 g
c. 11.69 g
d. 29.20 g
e. 58.40 g

32. The molecular weight of Sodium Bicarbonate BP is 84. How many grams of Sodium Bicarbonate BP would be required to produce 75 mL of 1 mmol/mL solution?

a. 0.63 g
b. 4.2 g
c. 6.3 g
d. 8.4 g
e. 12.6 g

Formulation questions

This section contains details of extemporaneous products to be made in the same way as the examples earlier in this chapter. For each example, provide answers using the following sections:

1. Use of the product
2. Is it safe and suitable for the intended purpose?
3. Calculation of formula for preparation
4. Method of preparation
a. Solubility where applicable
b. Vehicle/diluent
c. Preservative
d. Flavouring when appropriate
5. Choice of container
6. Labelling considerations
a. Title
b. Quantitative particulars
c. Product-specific cautions (or additional labelling requirements)
d. Directions to patient – interpretation of Latin abbreviations where necessary
e. Recommended *British National Formulary* cautions when suitable
f. Discard date
g. Sample label (you can assume that the name and address of the pharmacy and the words 'Keep out of the reach and sight of children' are pre-printed on the label)
7. Advice to patient

33. You receive a prescription in your pharmacy with the following details.

Patient:	Miss Julie Jordan, 21 Fair View, Astonbury
Age:	2 months
Prescription:	Sodium Bicarbonate Solution 0.5 mmol/mL
Directions:	10 mL bd with feeds
Mitte:	200 mL

34. You receive a prescription in your pharmacy with the following details.

Patient:	Mrs Sally Burns, 14 Netherton Grove, Astonbury
Age:	45
Prescription:	Hibitane solution 0.05%
Directions:	For cleansing the affected area
Mitte:	150 mL

Overview

Upon completion of this chapter, you should be able to:

- understand the difference between diffusible and indiffusible suspensions
- prepare diffusible and indiffusible suspensions from first principles
- select the correct amount of suspending agent to use when preparing an indiffusible suspension
- understand the 'doubling-up' technique
- select an appropriate container in which to package a suspension
- prepare an appropriate label for a suspension.

Introduction and overview of suspensions

Suspensions are important pharmaceutical dosage forms that are still widely in use today. Owing to their versatility they are often used in situations where an 'emergency' formulation is required.

Common pharmaceutical products that are suspensions include:

- ear drops
- enemas
- inhalations
- lotions
- mixtures for oral use.

British Pharmacopoeia **(BP) definition (oral suspensions)**

Oral suspensions are oral liquids containing one or more active ingredients suspended in a suitable vehicle. Suspended solids may slowly separate on standing but are easily redispersed.

Diffusible and indiffusible suspensions

Diffusible suspensions

These are suspensions containing light powders which are insoluble, or only very slightly soluble in the vehicle, but which on shaking disperse evenly throughout the vehicle for long enough to allow an accurate dose to be poured.

Definition

A pharmaceutical suspension is a preparation where at least one of the active ingredients is suspended throughout the vehicle. In contrast to solutions (see Chapter 2), in a suspension at least one of the ingredients is not dissolved in the vehicle and so the preparation will require shaking before a dose is administered.

KeyPoints

Advantages and disadvantages of suspensions as dosage forms

Advantages
- Insoluble drugs may be more palatable.
- Insoluble drugs may be more stable.
- Suspended insoluble powders are easy to swallow.
- The suspension format enables easy administration of bulk insoluble powders.
- Absorption will be quicker than solid dosage forms.
- Lotions will leave a cooling layer of medicament on the skin.
- It is theoretically possible to formulate sustained-release preparations.

Disadvantages
- Preparation requires shaking before use.
- Accuracy of dose is likely to be less than with equivalent solution.
- Storage conditions can affect disperse system.
- Suspensions are bulky, difficult to transport and prone to container breakages.

KeyPoints

- 0.2 g Tragacanth BP powder per 100 mL suspension
- 2 g Compound Tragacanth Powder BP per 100 mL suspension
- 2–3% Bentonite BP

Indiffusible suspensions

These are suspensions containing heavy powders which are insoluble in the vehicle and which on shaking do not disperse evenly throughout the vehicle long enough to allow an accurate dose to be poured.

In the preparation of indiffusible suspensions, the main difference when compared to diffusible suspensions is that the vehicle must be thickened to slow down the rate at which the powder settles. This is achieved by the addition of a suspending agent.

Choice of suspending agent

The amount of suspending agent used in any given formulation depends on the volume of vehicle being thickened. It does not vary with the amount of powder in the preparation. A suspending agent is intended to increase the viscosity of the vehicle and therefore slow down sedimentation rates. This outcome could also be achieved by decreasing the particle size of the powder in suspension.

The most common suspending agents used in extemporaneous dispensing are Tragacanth BP (internal or external suspensions), Compound Tragacanth Powder BP (containing: 15% Tragacanth BP, 20% Acacia BP, 20% Starch BP and 45% Sucrose BP) (internal suspensions) and Bentonite BP (external suspensions). Details on the appropriate quantities to use can be found below.

General method

General method for the preparation of a suspension containing a diffusible solid

1. Check the solubility in the vehicle of all solids in the mixture.
2. Calculate the quantities of vehicle required to dissolve any soluble solids.
3. Prepare any Double Strength Chloroform Water BP required.
4. Weigh all solids on a Class II or electronic balance.

5. Dissolve all soluble solids in the vehicle in a small glass beaker using the same procedures as outlined in Chapter 2.
6. Mix any insoluble diffusible powders in a porcelain mortar using the 'doubling-up' technique to ensure complete mixing (see key point below).
7. Add a small quantity of the vehicle (which may or may not be a solution of the soluble ingredients) to the solids in the mortar and mix using a pestle to form a smooth paste.
8. Add further vehicle in small quantities, and continue mixing until the mixture in the mortar is of a pourable consistency.
9. Transfer the contents of the mortar to a conical measure of suitable size.
10. Rinse out the mortar with more vehicle and add any rinsings to the conical measure.
11. Add remaining liquid ingredients to the mixture in the conical measure. (These are added now, as some may be volatile and therefore exposure while mixing needs to be reduced to prevent loss of the ingredient by evaporation.)
12. Make up to final volume with vehicle.
13. Stir gently, transfer to a suitable container, ensuring that all the solid is transferred from the conical measure to the bottle, and label ready to be dispensed to the patient.

See Suspensions video for a demonstration of the preparation of a diffusible suspension.

General method for the preparation of a suspension containing an indiffusible solid

Oral indiffusible suspensions are prepared using the same basic principles as for oral diffusible suspensions. The main difference is that the preparation will require the addition of a suspending agent. The suspending agent of choice will normally be combined with the indiffusible solid using the 'doubling-up' technique before incorporation into the product.

1. Check the solubility in the vehicle of all solids in the mixture.

Tips

Alternatively, the contents of the mortar could be transferred directly to a pre-prepared tared container. Rinsings from the mortar and other liquid ingredients could then be added to the bottle before making up to final volume. This would prevent any possible transference loss caused by powders sedimenting in the conical measure.

KeyPoints

The 'doubling-up' technique
1. Weigh the powder present in the smallest volume (powder A) and place in the mortar.
2. Weigh the powder present in the next largest volume (powder B) and place on labelled weighing paper.
3. Add approximately the same amount of powder B as powder A in the mortar.
4. Mix well with pestle.
5. Continue adding an amount of powder B that is approximately the same as that in the mortar and mix with the pestle, i.e. doubling the amount of powder in the mortar at each addition.
6. If further powders are to be added, add these in increasing order of volume as in parts 3, 4 and 5 above.

See Powders video for a demonstration of the 'doubling-up' technique.

2. Calculate the quantities of vehicle required to dissolve any soluble solids.
3. Prepare any Double Strength Chloroform Water BP required.
4. Weigh all solids on a Class II or electronic balance.
5. Dissolve all soluble solids in the vehicle in a small glass beaker.
6. Mix any insoluble indiffusible powders and the suspending agent in a porcelain mortar using the 'doubling-up' technique to ensure complete mixing.
7. Add a small quantity of the vehicle (which may or may not be a solution of the soluble ingredients) to the solids in the mortar and mix using a pestle to form a smooth paste.
8. Add further vehicle in small quantities, and continue mixing until the mixture in the mortar is a pourable consistency.
9. Transfer the contents of the mortar to a conical measure of suitable size.
10. Rinse out the mortar with more vehicle and add any rinsings to the conical measure.
11. Add remaining liquid ingredients to the mixture in the conical measure. (These are added now, as some may be volatile and therefore exposure while mixing needs to be reduced to prevent loss of the ingredient by evaporation.)
12. Make up to final volume with vehicle.
13. Stir gently, transfer to a suitable container, ensuring that all the solid is transferred from the conical measure to the bottle, and label ready to be dispensed to the patient.

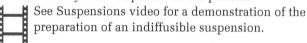 See Suspensions video for a demonstration of the preparation of an indiffusible suspension.

Worked examples

Example 3.1
The preparation of Magnesium Trisilicate Mixture BP
You receive a prescription in your pharmacy with the following details:

Patient:	Mr Peter Burley, 74 Stone Lane, Astonbury
Age:	37
Prescription:	Mist Mag Trisil
Directions:	10 mL tds ex aqua
Mitte:	150 mL

1. **Use of the product**
   Used to treat indigestion/dyspepsia (*British National Formulary* 61st edn, p 45).

2. **Is it safe and suitable for the intended purpose?**
    This is an official preparation, therefore the formula is safe
    and suitable for purpose. The dose of 10 mL three times a day
    is consistent with the recommended dosage (*British National
    Formulary* 61st edn, p 45).
3. **Calculation of formula for preparation**
    Prepare 150 mL of Magnesium Trisilicate Mixture BP.

Product formula
(from the *British Pharmacopoeia* 2007, p 2725)

	Master	100 mL	50 mL	150 mL
Magnesium Trisilicate BP	50 g	5 g	2.5 g	7.5 g
Light Magnesium Carbonate BP	50 g	5 g	2.5 g	7.5 g
Sodium Bicarbonate BP	50 g	5 g	2.5 g	7.5 g
Concentrated Peppermint Emulsion BP	25 mL	2.5 mL	1.25 mL	3.75 mL
Double Strength Chloroform Water BP	500 mL	50 mL	25 mL	75 mL
Potable water	to 1000 mL	to 100 mL	to 50 mL	to 150 mL

Interim formula for Double Strength Chloroform Water BP

Concentrated Chloroform Water BPC 1959	5 mL
Potable water	to 100 mL

4. **Method of preparation**
a. Solubility where applicable
    Sodium Bicarbonate BP is soluble 1 in 11 in water (*British
    Pharmacopoeia* 1988, p 509). Therefore to dissolve 7.5 g
    Sodium Bicarbonate BP, a minimum of 7.5 × 11 = 82.5 mL of
    water would be required. As this is greater than 50% of the
    mixture, a solution of Double Strength Chloroform Water BP
    and water would be used for dissolution.
b. Vehicle/diluent
    Double Strength Chloroform Water BP and potable water
    would be used as the vehicle as per the product formula.
c. Preservative
    Double Strength Chloroform Water BP is included in this
    product as the preservative as per the product formula.
d. Flavouring when appropriate
    Concentrated Peppermint Emulsion BP is included in this
    product as the flavouring as per the product formula.

    The following method would be used to prepare 150 mL of
    Magnesium Trisilicate Mixture BP from the formula above:

Tips

1. Using the master formula from the *British Pharmacopoeia* for 1000 mL of final product, calculate the quantity of ingredients required to produce the final volume needed (150 mL).

2. Calculate the composition of a convenient quantity of Double Strength Chloroform Water BP, sufficient to satisfy the formula requirements but also enabling simple, accurate measurement of the concentrated component.

Method of compounding for Double Strength Chloroform Water BP

a. In this case, 75 mL of Double Strength Chloroform Water BP is required and so it would be sensible to prepare 100 mL. To prepare 100 mL Double Strength Chloroform Water BP, measure 5 mL of Concentrated Chloroform Water BPC 1959 accurately using a 5 mL conical measure.

b. Add approximately 90 mL of potable water to a 100 mL conical measure (i.e. sufficient water to enable dissolution of the concentrated chloroform component without reaching the final volume of the product).

c. Add the measured Concentrated Chloroform Water BPC 1959 to the water in the conical measure.

d. Stir gently and then accurately make up to volume with potable water.

e. Visually check that no undissolved chloroform remains at the bottom of the measure.

Noting that Sodium Bicarbonate BP is soluble 1 in 11 with water, a minimum of 11 mL of water would be required to dissolve 1 g of Sodium Bicarbonate BP.

The final volume of Magnesium Trisilicate Mixture BP required (150 mL) will contain 7.5 g of Sodium Bicarbonate BP. As 1 g of sodium bicarbonate is soluble in 11 mL, 7.5 g is soluble in 82.5 mL (7.5 × 11 = 82.5 mL).

Therefore a minimum of 82.5 mL of vehicle would be required to dissolve the 7.5 g of sodium bicarbonate in this example. For ease of compounding, choose a convenient volume of vehicle, say 90 mL, in which to dissolve the solute initially. When choosing the amount of vehicle to use for dissolution, it is important to consider the total amount of each liquid ingredient in the preparation to ensure that only the correct amounts are added or the final product does not go over volume.

3. Weigh 7.5 g Magnesium Trisilicate BP on a Class II or electronic balance.
4. Weigh 7.5 g Light Magnesium Carbonate BP on a Class II or electronic balance.
5. Weigh 7.5 g Sodium Bicarbonate BP on a Class II or electronic balance.
6. Measure 3.75 mL Concentrated Peppermint Emulsion BP using a 1 mL and 5 mL syringe.
7. Accurately measure 75 mL Double Strength Chloroform Water BP using a 100 mL measure. To this add approximately 15 mL potable water in order to produce 90 mL of vehicle which should be poured into a beaker (in order to produce sufficient volume to dissolve the 7.5 g Sodium Bicarbonate BP).
8. The Sodium Bicarbonate BP (7.5 g) should be added to the vehicle, thus following the principle of adding solutes to solvents.
9. Stir to aid dissolution.
10. Transfer the Magnesium Trisilicate BP to a porcelain mortar.
11. Add the Light Magnesium Carbonate BP to the Magnesium Trisilicate BP in the mortar using the 'doubling-up' technique and stir with a pestle to ensure even mixing.
12. Add a small amount of the sodium bicarbonate solution to the powder in the mortar and mix with a pestle to make a smooth paste.
13. Slowly continue adding the sodium bicarbonate solution until the paste is pourable.
14. Transfer the contents of the mortar to a 250 mL conical measure.
15. Rinse out the mortar with more sodium bicarbonate solution and add the rinsings to the conical measure.
16. Add the Concentrated Peppermint Emulsion BP to the mixture in the conical measure.

Tips

As discussed above, in this example 90 mL of vehicle is required to dissolve the Sodium Bicarbonate BP. It is important to consider the total amount of each liquid ingredient in the product to ensure that only the correct amounts are added.

In this example, it would be incorrect to dissolve the Sodium Bicarbonate BP in 90 mL of Double Strength Chloroform Water BP as the final volume of the preparation only contains 75 mL. Equally, it would also be incorrect to dissolve the Sodium Bicarbonate BP in 90 mL of water as the final volume of the preparation will contain less than 75 mL.

In this case, all the Double Strength Chloroform Water BP is used (75 mL) along with enough potable water to reach the desired volume (approximately 15 mL).

Tips

The Magnesium Trisilicate BP is added to the mortar first as, although the weights of the insoluble solids are identical, the volume occupied by the powers differs markedly. The Magnesium Trisilicate BP occupies the smallest volume and therefore is the first powder to be added to the mortar.

17. Make up to volume with any remaining solution and potable water.
18. Transfer the solution to a 150 mL amber flat medical bottle with a child-resistant closure and label.

5. **Choice of container**

   A plain amber bottle with a child-resistant closure would be most suitable as the preparation is a suspension for internal use.

6. **Labelling considerations**

   a. Title

      The product is official, therefore the following title would be suitable: 'Magnesium Trisilicate Mixture BP'.

   b. Quantitative particulars

      Quantitative particulars are not required as the product is official.

   c. Product-specific cautions (or additional labelling requirements)

      'Shake the bottle' will need to be added to the label as the product is a suspension and will need shaking before use to ensure an accurate dose is measured.

   d. Directions to patient – interpretation of Latin abbreviations where necessary

      'Take TWO 5 mL spoonfuls THREE times a day in water.'

   e. Recommended *British National Formulary* cautions when suitable

      Not applicable.

   f. Discard date

      The *British Pharmacopoeia* states that this product should be recently prepared, therefore it will attract a 4-week discard date (*British Pharmacopoeia* 2004, p 2561).

   g. Sample label (you can assume that the name and address of the pharmacy and the words 'Keep out of the reach and sight of children' are pre-printed on the label):

Magnesium Trisilicate Mixture BP   150 mL
Take TWO 5 mL spoonfuls THREE times a day in water.
Shake the bottle
Do not use after (4 weeks)
Mr Peter Burley   Date of dispensing

---

7. **Advice to patient**

   The patient would be advised to mix two 5 mL spoonfuls with an equal volume of water and take three times a day. In addition, the discard date and the need to shake the bottle before a dose is measured would be highlighted.

Example 3.2
The preparation of Paediatric Chalk Mixture BP
You receive a prescription in your pharmacy with the following details:

Patient:	Miss Jane Peacock, 12a The Ridings, Astonbury
Age:	7
Prescription:	Chalk Mixture Paed
Directions:	10 mL qds
Mitte:	100 mL

1. **Use of the product**
   Used to treat diarrhoea (*Martindale* 26th edn, p 131).
2. **Is it safe and suitable for the intended purpose?**
   This is an official preparation, therefore the formula is safe and suitable for purpose. The dose of 10 mL four times a day is consistent with the recommended dosage of 10 mL (*Martindale* 26th edn, p 132) three to four times a day by convention.
3. **Calculation of formula for preparation**
   Prepare 100 mL of Paediatric Chalk Mixture BP.

Product formula
(from the *British Pharmacopoeia* 1988, p 724)

	Master	100 mL
Chalk BP	20 g	2 g
Tragacanth BP	2 g	200 mg
Concentrated Cinnamon Water BP	4 mL	0.4 mL
Syrup BP	100 mL	10 mL
Double Strength Chloroform Water BP	500 mL	50 mL
Potable water	to 1000 mL	to 100 mL

Interim formula for Double Strength Chloroform Water

Concentrated Chloroform Water BPC 1959	2.5 mL
Potable water	to 50 mL

4. **Method of preparation**
a. Solubility where applicable
   Not applicable as there are no soluble solids in the preparation.
b. Vehicle/diluent
   Double Strength Chloroform Water BP and potable water would be used as the vehicle as per the product formula.
c. Preservative
   Double Strength Chloroform Water BP is included in this product as the preservative as per the product formula.

    d. Flavouring when appropriate
Concentrated Cinnamon Water BP is included in this product as the flavouring as per the product formula.

The following method would be used to prepare 100 mL of Paediatric Chalk Mixture BP from the formula above:

1. Calculate the composition of a convenient quantity of Double Strength Chloroform Water BP, sufficient to satisfy the formula requirements but also enabling simple, accurate measurement of the concentrated component.

Method of compounding for Double Strength Chloroform Water BP

    a. In this case, 50 mL of Double Strength Chloroform Water BP is required and so it would be sensible to prepare 50 mL. To prepare 50 mL Double Strength Chloroform Water BP, measure 2.5 mL of Concentrated Chloroform water BPC 1959 accurately using a 5 mL and a 1 mL syringe.

    b. Add approximately 45 mL of potable water to a 50 mL conical measure (i.e. sufficient water to enable dissolution of the concentrated chloroform component without reaching the final volume of the product).

    c. Add the measured Concentrated Chloroform Water BPC 1959 to the water in the conical measure.

    d. Stir gently and then accurately make up to volume with potable water.

    e. Visually check that no undissolved chloroform remains at the bottom of the measure.

2. Weigh 200 mg Tragacanth BP accurately on a Class II or electronic balance.
3. Weigh 2 g Chalk BP accurately on a Class II or electronic balance.
4. Measure 10 mL Syrup BP in a 10 mL conical measure.
5. Measure 0.4 mL Concentrated Cinnamon Water BP using a 1 mL syringe.
6. Measure 50 mL of Double Strength Chloroform Water BP in a 50 mL conical measure.
7. Transfer the Tragacanth BP to a porcelain mortar.
8. Add the Chalk BP to the mortar using the 'doubling-up' technique to mix the two powders.

Tips

Although it is a volatile ingredient, it is not bad practice to measure the Concentrated Cinnamon Water BP at this stage of the method as loss by evaporation will be avoided by measuring in a syringe.

It must also be remembered that, when using a syringe to measure ingredients, consideration must be given to the properties of the liquid being measured. This will avoid dissolution of the volume markings from the outside of the syringe, or even parts of the syringe itself, which may occur with some ingredients.

9. Add the Syrup BP to the mortar and mix to form a smooth paste.
10. Add some of the Double Strength Chloroform Water BP to the paste and mix until pourable.
11. Transfer the contents to a 100 mL conical measure.
12. Rinse out the mortar with more Double Strength Chloroform Water BP or potable water and add the rinsings to the conical measure.
13. Add the Concentrated Cinnamon Water BP to the mixture in the conical measure.
14. Make up to volume with any remaining Double Strength Chloroform Water BP and potable water.
15. Transfer to an amber flat medical bottle label and dispense.

Tips

The Tragacanth BP is included in the mixture because Chalk BP is an indiffusible solid and therefore it is necessary to add a suspending agent. They are admixed by the 'doubling-up' technique to ensure even mixing and therefore the successful suspension of the indiffusible chalk.

Tips

The Concentrated Cinnamon Water BP is the last ingredient to be added prior to making up to volume because it is a volatile ingredient.

5. **Choice of container**
   A plain amber bottle with a child-resistant closure would be most suitable as the preparation is a suspension for internal use.

Tips

Alternatively a bottle could be tared and the mixture made up to volume in the bottle.

6. **Labelling considerations**
a. Title
   The product is official, therefore the following title would be suitable: 'Paediatric Chalk Mixture BP'.
b. Quantitative particulars
   Quantitative particulars are not required as the product is official.
c. Product-specific cautions (or additional labelling requirements)
   'Shake the bottle' will need to be added to the label as the product is a suspension and will need shaking before use to ensure an accurate dose is measured.
d. Directions to patient – interpretation of Latin abbreviations where necessary
   'Give TWO 5 mL spoonfuls FOUR times a day'. The word 'give' is used here as the preparation is for a child and so will be administered by a parent or guardian.
e. Recommended *British National Formulary* cautions when suitable
   Not applicable.

f. Discard date
The *British Pharmacopoeia* states that this product should be recently prepared, therefore it will attract a 4-week discard date (*British Pharmacopoeia* 1988, p 724).

g. Sample label (you can assume that the name and address of the pharmacy and the words 'Keep out of the reach and sight of children' are pre-printed on the label):

Paediatric Chalk Mixture BP	**100 mL**
Give TWO 5 mL spoonfuls FOUR times a day.	
Shake the bottle	
Do not use after (4 weeks)	
Miss Jane Peacock	Date of dispensing

7. **Advice to patient**
The parent or guardian would be advised to give two 5 mL spoonfuls four times a day. In addition, the discard date and the need to shake the bottle before a dose is measured would be highlighted.

Example 3.3
The preparation of Menthol and Eucalyptus Inhalation BP

You receive a prescription in your pharmacy with the following details:

Patient:	Mr Adrian Temple, 45 High Street, Astonbury
Age:	29
Prescription:	Menthol and Eucalyptus Inhalation
Directions:	Use when required
Mitte:	50 mL

1. **Use of the product**
Used to relieve symptoms of bronchitis and nasal obstruction in acute rhinitis or sinusitis (*British National Formulary* 61st edn, p 204).

2. **Is it safe and suitable for the intended purpose?**
This is an official preparation, therefore the formula is safe and suitable for purpose. The *British National Formulary* suggests adding one teaspoonful to a pint of hot, not boiling, water and inhaling the vapour (*British National Formulary* 61st edn, p 204).

3. **Calculation of formula for preparation**
Prepare 50 mL of Menthol and Eucalyptus Inhalation BP.

Product formula
(from the *British Pharmacopoeia* 1980, p 577)

	Master	100 mL	50 mL
Menthol BP	20 g	2 g	1 g
Eucalyptus Oil BP	100 mL	10 mL	5 mL
Light Magnesium Carbonate BP	70 g	7 g	3.5 g
Potable water	to 1000 mL	to 100 mL	to 50 mL

4. Method of preparation

a. Solubility where applicable

The *British Pharmacopoeia* states that Menthol BP is freely soluble in fixed oils and liquid paraffin (*British Pharmacopoeia* 1988, p 358).

b. Vehicle/diluent

Potable water would be used as the vehicle as per the formula.

c. Preservative

No preservative is included in this product as per the product formula.

d. Flavouring when appropriate

This product is for external use and so no flavouring is required.

The following method would be used to prepare 50 mL of Menthol and Eucalyptus Inhalation BP from the formula above:

1. Weigh 1 g Menthol BP on a Class II or electronic balance.
2. Transfer to a glass mortar.
3. Measure 5 mL Eucalyptus Oil BP in a 5 mL conical measure.
4. Transfer to the mortar and mix with a pestle to dissolve the menthol.
5. Weigh 3.5 g Light Magnesium Carbonate BP on a Class II or electronic balance.
6. Add to the mortar and mix well.
7. Add potable water to the mixture in the mortar to form a pourable suspension.
8. Transfer the contents of the mortar to a 50 mL conical measure.
9. Rinse the mortar with potable water and add the rinsings to the conical measure.
10. Make up to volume with potable water.
11. Transfer to a 50 mL amber fluted bottle with a child-resistant closure and label.

Tips

Menthol BP is freely soluble in fixed and volatile oils.
The Eucalyptus Oil BP should be measured in a glass conical measure as the oil will dissolve away the graduation markings on a syringe.

5. **Choice of container**
    An amber fluted bottle with a child-resistant closure would be most suitable as the preparation is a suspension for external use.

6. **Labelling considerations**

a. Title
    The product is official, therefore the following title would be suitable: 'Menthol and Eucalyptus Inhalation BP'.

b. Quantitative particulars
    Quantitative particulars are not required as the product is official.

c. Product-specific cautions (or additional labelling requirements)
    'Shake the bottle' will need to be added to the label as the product is a suspension and will need shaking before use to ensure an accurate dose is measured.
    'Not to be taken' will need to be added to the label as the produce is an inhalation.

d. Directions to patient – interpretation of Latin abbreviations where necessary
    'Add ONE teaspoonful to a pint of hot, not boiling, water and inhale the vapour when required.'

e. Recommended *British National Formulary* cautions when suitable
    Not applicable.

f. Discard date
    The product is an inhalation and therefore it will attract a 4-week discard date.

g. Sample label (you can assume that the name and address of the pharmacy and the words 'Keep out of the reach and sight of children' are pre-printed on the label):

Tips

The directions to the patient will suggest the use of a teaspoon, rather than a 5 mL spoon. This is because a 5 mL spoon is supplied to a patient to measure a dose for internal administration. If a 5 mL spoon was supplied here, the patient might mistakenly think that the product was for internal administration.

Menthol and Eucalyptus Inhalation BP	50 mL
Add ONE teaspoonful to a pint of hot, not boiling, water and inhale the vapour when required.	
Shake the bottle	
Not to be taken	
Do not use after (4 weeks)	
Mr Adrian Temple	Date of dispensing

7. **Advice to patient**

The patient would be advised to add one teaspoonful to a pint of hot, not boiling, water and inhale the vapour when required. In addition, the discard date, the need to shake the bottle before a dose is measured and the fact that the product is not to be taken would be highlighted.

Example 3.4
The preparation of a magistral formulation from a hospital formula

You receive the following prescription for clobazam liquid:

Patient:	Mr Jim Smith, 34 Beaches Avenue, Astonbury
Age:	70
Prescription:	Clobazam liquid
Directions:	10 mg tds
Mitte:	10/7

1. **Use of the product**

Used to treat epilepsy and anxiety (*British National Formulary* 61st edn, p 293). In this case, you are informed by the patient that the product is being used to treat anxiety.

2. **Is it safe and suitable for the intended purpose?**

This is an unofficial preparation, therefore the formula will need to be checked to ensure that it is safe and suitable for purpose. The *British National Formulary* states that the usual adult dose to treat epilepsy is 20–30 mg daily; maximum 60 mg daily, and for anxiety is 20–30 mg daily in divided doses or as a single dose at bedtime, increased in severe anxiety (in hospital patients) to a maximum of 60 mg daily in divided doses; elderly (or debilitated) 10–20 mg daily (*British National Formulary* 61st edn, p 293). Therefore, a dose of 10 mg three times a day would seem appropriate.

3. **Calculation of formula for preparation**

Clobazam is only available commercially as 10 mg tablets. The patient is required to take 10 mg at each dose but unfortunately cannot swallow solid-dose preparations. The hospital pharmacy gives you the formula that has been used while the patient was in the hospital:

Tabs qs Clobazam	10 mg
Concentrated Peppermint Water BP	2% v/v
Glycerol BP	6% v/v
Syrup BP	25% v/v
Suspending agent	2% w/v
Freshly boiled and cooled purified water	to 100%

The patient needs to take 10 mg at each dose and the prescriber wants 30 doses (the dose is one to be taken three times a day and the prescriber has requested a 10-day supply).

Therefore if we make the formula as above, such that each 5 mL contains 10 mg clobazam, we would need to prepare 150 mL of suspension.

Prepare 150 mL clobazam suspension 10 mg/5 mL.

Product formula

	5 mL	50 mL	150 mL
Clobazam	10 mg	100 mg	300 mg
	1 tablet	10 tablets	30 tablets
Concentrated Peppermint Water BP	0.1 mL	1 mL	3 mL
Glycerol BP	0.3 mL	3 mL	9 mL
Syrup BP	1.25 mL	12.5 mL	37.5 mL
Compound Tragacanth Powder BP	100 mg	1 g	3 g
Freshly boiled and cooled purified water	to 5 mL	to 50 mL	to 150 mL

4.   Method of preparation

a.   Solubility where applicable

There are no soluble solids in this preparation; however, a suspending agent will need to be added as the product will be an indiffusible suspension.

b.   Vehicle/diluent

Freshly boiled and cooled purified water is used as the vehicle as no preservative is included in the preparation.

c.   Preservative

The product does not contain a preservative.

d.   Flavouring when appropriate

Concentrated Peppermint Water BP has been chosen as a flavouring to mask the taste of the suspension.

Tips

As the quantity of suspending agent required was indicated to be 2%, Compound Tragacanth Powder BP was chosen as the suspending agent as this is a suitable quantity to include to produce a reasonable suspension. Tragacanth BP itself is normally used in concentrations of 0.2%; therefore if Tragacanth BP had been used in a 2% strength, the suspension produced would have been unacceptable and more closely related to a solid dosage form than a liquid dosage form.

The following method would be used to prepare 150 mL of suspension using the formula above:

1.   Count out 30 clobazam tablets.
2.   Weigh 3 g Compound Tragacanth Powder BP using a Class II or electronic balance.
3.   Measure 9 mL of Glycerol BP using a 10 mL conical measure.
4.   Measure 37.5 mL Syrup BP using a 37 mL measure and a suitably graduated 5 mL syringe
5.   Transfer the tablets to a glass mortar and grind them to make a smooth powder.

6. Transfer the powder to a porcelain mortar and add the Compound Tragacanth Powder BP using the 'doubling-up' technique. Mix with the pestle.
7. Add the Glycerol BP to the powders in the mortar and mix to make a paste.
8. Add the Syrup BP to the mortar to make a pourable paste.
9. Transfer the contents of the mortar to a 250 mL conical measure.
10. Rinse out the mortar with more syrup and freshly boiled and cooled purified water.
11. Add rinsings to the mixture in the conical measure.
12. Measure 3 mL Concentrated Peppermint Water BP using a 5 mL conical measure.
13. Add the Concentrated Peppermint Water BP to the mixture in the conical measure.
14. Make up to volume with freshly boiled and cooled purified water.
15. Stir and transfer to a 150 mL amber flat medical bottle with a child-resistant closure, label and dispense to the patient.

5. **Choice of container**
   A plain amber bottle with a child-resistant closure would be most suitable as the preparation is a suspension for internal use.
6. **Labelling considerations**
   a. Title
      The product is unofficial, therefore the following title would be suitable: 'Clobazam 10 mg/5 mL suspension'.
   b. Quantitative particulars
      Quantitative particulars are required, as the product is unofficial. As this is a product for internal use, the quantitative particulars will be expressed per dose (i.e. per 5 mL).

Each 5 mL contains:	
Clobazam	10 mg
Concentrated Peppermint Water BP	0.1 mL
Glycerol BP	0.3 mL
Syrup BP	1.25 mL
Compound Tragacanth Powder BP	100 mg
Freshly boiled and cooled purified water	to 5 mL

   c. Product-specific cautions (or additional labelling requirements)
      'Shake the bottle' will need to be added to the label as the product is a suspension and will need shaking before use to ensure an accurate dose is measured.

d. Directions to patient – interpretation of Latin abbreviations where necessary

'Take ONE 5 mL spoonful THREE times a day.'

e. Recommended *British National Formulary* cautions when suitable

The *British National Formulary* recommends the following cautions (*British National Formulary* 61st edn, p 293):

Label 2 – 'Warning: This medicine may make you sleepy. If this happens, do not drive or use tools or machines. Do not drink alcohol.'

or

Label 19 – 'Warning: This medicine makes you sleepy. If you still feel sleepy the next day, do not drive or use tools or machines. Do not drink alcohol.'

and

Label 8 – 'Warning: Do not stop taking this medicine unless your doctor tells you to stop.'

As this preparation is being used to treat anxiety, label 2 would be most suitable. Label 8 would only be required if the product was being used to treat epilepsy.

f. Discard date.

As the product is unofficial and doesn't contain a preservative, a 2-week discard would be most suitable.

g. Sample label (you can assume that the name and address of the pharmacy and the words 'Keep out of the reach and sight of children' are pre-printed on the label):

Clobazam 10 mg/5 mL suspension	**150 mL**

Take ONE 5 mL spoonful THREE times a day.
Shake the bottle

Warning: This medicine may make you sleepy. If this happens, do not drive or use tools or machines. Do not drink alcohol.

Do not use after (2 weeks)

Each 5 mL contains:

Clobazam	10 mg
Concentrated Peppermint Water BP	0.1 mL
Glycerol BP	0.3 mL
Syrup BP	1.25 mL
Compound Tragacanth Powder BP	100 mg
Freshly boiled and cooled purified water	to 5 mL

Mr Jim Smith	Date of dispensing

7. **Advice to patient**

The patient would be advised to take one 5 mL spoonful three times a day. In addition, the discard date, the need to shake the bottle and additional *British National Formulary* warnings would be highlighted.

Self-assessment

1. Distinguish between 'diffusible' and 'indiffusible' solids.

2. Which of the following is an example of an indiffusible powder?
a. Chalk BP
b. Light Kaolin BP
c. Light Magnesium Carbonate BP
d. Sodium Bicarbonate BP

3. Which of the following is an example of a diffusible powder?
a. Calamine BP
b. Lactose BP
c. Sodium Bicarbonate BP
d. Magnesium Trisilicate BP

4. Which of the following suspending agents would be unsuitable for use for a suspension intended for the oral route of administration?
a. Tragacanth BP
b. Compound Tragacanth Powder BP
c. Bentonite BP
d. Methylcellulose BP

5. Which of the following is not a suspension?
a. Paediatric Kaolin Mixture BP
b. Zinc Sulphate Lotion BP
c. Sulphur Lotion Compound BPC 1973
d. Calpol

6. What is the formula for Compound Tragacanth Powder BP?

7. Why is Compound Tragacanth Powder BP unsuitable for external use?

8. What specific instruction should be included on the label of all suspensions?

9. You are asked to prepare 100 mL of an unofficial suspension with a dose statement 15 mL tds ex aq. How would the quantitative particulars be expressed on a dispensed type label?
a. Amount per 5 mL
b. Amount per 10 mL
c. Amount per 15 mL
d. Amount per 100 mL

10. You have been given the following prescription: 'Paracetamol suspension 1 g po qds'. In the dispensary you have a suspension that contains 120 mg of paracetamol in each 5 mL spoonful. How much suspension would you need to dispense, to the nearest 100 mL, for a 2-week supply?
   a. 1.1 litres
   b. 1.5 litres
   c. 2.0 litres
   d. 2.3 litres
   e. 2.4 litres

11. How much Tragacanth BP would be required to make 200 mL of an indiffusible suspension?
   a. 100 mg
   b. 200 mg
   c. 400 mg
   d. 600 mg
   e. 4 g

12. A prescriber sends you a prescription asking for 100 mL of a lotion containing Calamine BP 15% and Zinc Oxide BP 5%. Suggest a suitable formula for this product and the reasons for your choices.

13. You are asked to prepare 150 mL of a suspension containing an indiffusible solid. How much Compound Tragacanth Powder BP would need to be added to achieve an acceptable product?
   a. 200 mg
   b. 300 mg
   c. 2 g
   d. 3 g

14. Which of the following statements is true?
   a. Freshly boiled and cooled purified water must be used when making suspensions.
   b. All suspensions will attract a 4-week discard date.
   c. All suspensions need a direction to shake the bottle.
   d. Suspensions containing a suspending agent do not need a direction to shake the bottle.

15. Which of the following statements is false?
   a. The amount of suspending agent used depends on the volume of the suspension.
   b. The amount of suspending agent used depends on the amount of powder to be suspended.
   c. Diffusible suspensions contain an insoluble powder that is light and easily wettable.
   d. Adding a thickening agent to increase viscosity is the simplest way to ensure uniformity of dose in a suspension.

Formulation questions

This section contains details of extemporaneous products to be made in the same way as the examples earlier in this chapter. For each example, provide answers using the following sections:

1. Use of the product
2. Is it safe and suitable for the intended purpose?
3. Calculation of formula for preparation
4. Method of preparation
a. Solubility where applicable
b. Vehicle/diluent
c. Preservative
d. Flavouring when appropriate
5. Choice of container
6. Labelling considerations
a. Title
b. Quantitative particulars
c. Product-specific cautions (or additional labelling requirements)
d. Directions to patient – interpretation of Latin abbreviations where necessary
e. Recommended *British National Formulary* cautions when suitable
f. Discard date
g. Sample label (you can assume that the name and address of the pharmacy and the words 'Keep out of the reach and sight of children' are pre-printed on the label)
7. Advice to patient

16. You receive a prescription in your pharmacy with the following details:

Patient:	Mrs Sally Marlow, 12 George Street, Astonbury
Age:	34
Prescription:	Magnesium Carbonate Mixture BPC
Directions:	10 mL tds ex aq
Mitte:	150 mL

17. You receive a prescription in your pharmacy with the following details:

Patient:	Mr Edward Smith, 145 Oak Street, Astonbury	
Age:	57	
Prescription:	Ft Lotion	
	Zinc Oxide	25%
	Purified Talc	25%
	Glycerin	5%
	Suspending Agent	qs
	Water	to 100%
Directions:	Apply prn for itching	
Mitte:	50 mL	

chapter 4
Emulsions

Overview

Upon completion of this chapter, you should be able to:
- understand how to calculate the quantities of ingredients required to make a primary emulsion
- prepare an emulsion from first principles
- select an appropriate container in which to package an emulsion
- prepare an appropriate label for an emulsion.

Introduction and overview of emulsions

The pharmaceutical term 'emulsion' is solely used to describe preparations intended for internal use, i.e. via the oral route of administration. Emulsion formulations for external use are always given a different title that reflects their use, e.g. application, lotion and cream.

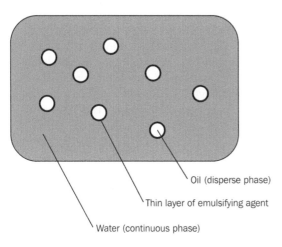

Oil (disperse phase)

Thin layer of emulsifying agent

Water (continuous phase)

Figure 4.1. Illustration of an oil-in-water emulsion.

British Pharmacopoeia (BP) definition (oral emulsions)
Oral emulsions are oral liquids containing one or more active ingredients. They are stabilised oil-in-water dispersions, either or both phases of which may contain dissolved solids. Solids may also be suspended in oral emulsions.

Definition

An emulsion is essentially a liquid preparation containing a mixture of oil and water that is rendered homogeneous by the addition of an emulsifying agent. The emulsifying agent ensures that the oil phase is finely dispersed throughout the water as minute globules (Figure 4.1). This type of emulsion is termed an 'oil-in-water' emulsion. The oily phase (disperse phase) is dispersed through the aqueous phase (continuous phase). Generally all oral dose emulsions tend to be oil-in-water as the oily phase is usually less pleasant to take and more difficult to flavour. 'Water-in-oil' emulsions can be formed but these tend to be those with external uses.

KeyPoints

Advantages and disadvantages of emulsions as dosage forms

Advantages
- Unpalatable oils can be administered in palatable form.
- Unpalatable oil-soluble drugs can be administered in palatable form.
- The aqueous phase is easily flavoured.
- The oily sensation is easily removed.
- The rate of absorption is increased.
- It is possible to include two incompatible ingredients, one in each phase of the emulsion.

Disadvantages
- Preparation needs to be shaken well before use.
- A measuring device is needed for administration.
- A degree of technical accuracy is needed to measure a dose.
- Storage conditions may affect stability.
- Bulky, difficult to transport and prone to container breakages.
- Liable to microbial contamination which can lead to cracking.

When issued for use, oral emulsions should be supplied in wide-mouthed bottles.

Extemporaneous preparation

In oral emulsions prepared according to the formula and directions given for extemporaneous preparation, the quantity of emulsifying agent specified in individual monographs may be reduced to yield a preparation of suitable consistency provided that by so doing the stability of the preparation is not adversely affected.

Stability of emulsions

Emulsions can break down in the following ways:
- cracking
- creaming
- phase inversion.

Cracking

This is the term applied when the disperse phase coalesces and forms a separate layer. Redispersion cannot be achieved by shaking and the preparation is no longer an emulsion. Cracking can occur if the oil turns rancid during storage. The acid formed denatures the emulsifying agent, causing the two phases to separate.

Creaming

In creaming, the oil separates out, forming a layer on top of the emulsion, but it usually remains in globules so that it can be redispersed on shaking (e.g. the cream on the top of a pint of milk). This is undesirable as the product appearance is poor and if the product is not adequately shaken there is a risk of the patient obtaining an incorrect dose. Creaming is less likely to occur if the viscosity of the continuous phase is increased.

Phase inversion

This is the process when an oil-in-water emulsion changes to a water-in-oil emulsion or vice versa. For stability of an emulsion, the optimum range of concentration of dispersed phase is 30–60% of the total volume. If the disperse phase exceeds this, the stability of the emulsion is questionable. As the concentration of the disperse phase approaches a theoretical maximum of 74% of the total volume, phase inversion is more likely to occur.

Summary of the problems encountered by emulsions

A summary of the problems encountered by emulsions is given in the key points box above.

General method

The theory of emulsification is based on the study of the most naturally occurring emulsion, milk. If examined closely, milk will be seen to consist of fatty globules, surrounded by a layer of casein, suspended in water. When a pharmaceutical emulsion is made, the principal considerations are the same. The object is to divide the oily phase completely into minute globules, surround each globule with an envelope of suspending agent (e.g. Acacia BP) and finally suspend the globules in the aqueous phase (Figure 4.1).

As with other liquid preparations for oral use, emulsions will have in the formulation a vehicle containing added flavouring or colourings as required. There is also the need for a preservative, which is usually chloroform, in the form of Double Strength Chloroform Water BP. In addition an emulsion will also need an emulsifying agent (or emulgent).

Continental and dry gum method

Although emulsions may be made by a variety of methods (for example, using methylcellulose gum in the preparation of Liquid Paraffin Emulsion BP: see Example 4.4 below), extemporaneously prepared emulsions for oral administration are usually made by the continental or dry gum method, where the emulsion is formed by mixing the emulsifying gum (usually Acacia BP) with the oil which is then mixed with the aqueous phase. The only differences between the continental and dry gum methods are the proportions of constituents within the primary emulsion (for example, fixed-oil emulsions made by

---

KeyPoints

Summary of the problems encountered by emulsions

Creaming
Separation of the emulsion into two regions, one containing more of the disperse phase.
Possible reasons for problem
- lack of stability of the system
- product not homogeneous.
Can the emulsion be saved?
The emulsion will reform on shaking.

Cracking
The globules of the disperse phase coalesce and there is separation of the disperse phase into a separate layer.
Possible reasons for problem:
- incompatible emulsifying agent
- decomposition of the emulsifying agent
- change of storage temperature.
Can the emulsion be saved?
The emulsion will not reform on shaking.

Phase inversion
From oil-in-water to water-in-oil or from water-in-oil to oil-in-water.
Possible reason for problem
- amount of disperse phase greater than 74%.
Can the emulsion be saved?
The emulsion will not reform on shaking.

the continental method would use a ratio of 4:3:2 rather than 4:2:1 with the dry gum method).

Internal emulsions prepared by the dry gum method should contain, in addition to the oil to be emulsified:

- a vehicle
  Freshly boiled and cooled purified water is normally used because of the increased risk from microbial contamination.
- a preservative
  This is usually added to the product as Double Strength Chloroform Water BP at 50% of the volume of the vehicle. If freshly boiled and cooled purified water is used as the vehicle, it would be appropriate to manufacture the Double Strength Chloroform Water BP using freshly boiled and cooled purified water rather than potable water.
- an emulsifying agent (or emulgent)
  The quantity of emulsifying agent added is determined by the type of oil to be emulsified and the quantity of emulsion to be prepared
- additional flavouring if required
- additional colouring if required.

Calculation of the amount of emulsifying agent to be used in the preparation of an emulsion

The amount of emulsifying agent used is dependent on the amount and type of oil to be emulsified. Oils can be divided into three categories: fixed oils, mineral oils and volatile oils.

Fixed oils
- Oil: 4 parts by volume
- Aqueous phase: 2 parts by volume
- Gum: 1 part by weight

Mineral oils
- Oil: 3 parts by volume
- Aqueous phase: 2 parts by volume
- Gum: 1 part by weight

Volatile (aromatic) oils
- Oil: 2 parts by volume
- Aqueous phase: 2 parts by volume
- Gum: 1 part by weight

These proportions are important when making the primary emulsion, to prevent the emulsion breaking down on dilution or storage.

The quantities for primary emulsions (in parts) are summarised in the key points box.

Wet gum method

The proportions of oil, water and emulsifying agent for the preparation of the primary emulsion are the same as those used in the dry gum method. The difference is in the method of preparation.

Using this method the acacia powder is added to the mortar and then triturated with the water until the gum is dissolved and a mucilage formed. The oil is then added to the mucilage drop by drop while triturating continuously. When nearly all the oil has been added the resulting mixture in the mortar may appear a little poor with some of the oil appearing to be absorbed. This can be rectified by the addition of slightly more water. The trituration continues until all the oil has been added, adding extra small amounts of water when necessary. When all the oil has been added triturate until a smooth primary emulsion is obtained.

In the main, this method has fallen out of favour as it takes much longer than the dry gum method. It should be noted that there is less chance of failure with this method provided the oil is added very slowly and in small quantities. It also means that the reasons for failure when using the dry gum method (outlined above) have been eliminated.

General method of preparation of an emulsion using the dry gum method

It is relatively easy for an emulsion to crack, resulting in a failed product. Remember that the key points above are critical when preparing emulsions.

The preparation of an emulsion has two main components:
1. preparation of a concentrate called the primary emulsion
2. dilution of the concentrate.

KeyPoints

The ratio of oily phase to aqueous phase to gum in a primary emulsion

Type of oil	Oil	Aqueous	Gum
Fixed	4	2	1
Mineral	3	2	1
Volatile	2	2	1

Tips

Accurate weighing and measuring of the components in the primary emulsion are important when making the primary emulsion to prevent the emulsion breaking down on storage or dilution.

KeyPoints

Clean, dry equipment
All equipment should be thoroughly cleaned, rinsed with water and carefully dried before use, particularly measures, mortars and pestles.

Accurate quantities
Accurate quantities are essential. Check weighing/measuring technique and minimise transference losses; for example, allow oil to drain from measure.

Have all ingredients ready
Correct rate of addition is important. Ingredients for the primary emulsion should all be weighed and measured before starting to make the product.

Preparation of the primary emulsion

1. Measure the oil accurately in a dry measure. Transfer the oil into a large dry porcelain mortar, allowing all the oil to drain out.
2. Measure the quantity of aqueous vehicle required for the primary emulsion. Place this within easy reach.
3. Weigh the emulsifying agent and place on the oil in the mortar. Mix lightly with the pestle, just sufficient to disperse any lumps. Caution: overmixing generates heat, which may denature the emulsifying agent and result in a poor product.
4. Add all of the required aqueous vehicle in one addition. Then mix vigorously, using the pestle with a shearing action in one direction.
5. When the product becomes white and produces a clicking sound, the primary emulsion has been formed. The product should be a thick, white cream. Increased degree of whiteness indicates a better-quality product. Oil globules or slicks should not be apparent.

Dilution of the primary emulsion

1. Dilute the primary emulsion drop by drop with very small volumes of the remaining aqueous vehicle. Mix carefully with the pestle in one direction.
2. Transfer emulsion to a measure, with rinsings. Add other liquid ingredients if necessary and make up to the final volume.

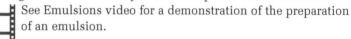 See Emulsions video for a demonstration of the preparation of an emulsion.

Worked examples

Example 4.1
The preparation of a magistral formulation from a doctor's prescription

You receive a prescription in your pharmacy with the following details:

Patient:	Mrs Fiona Archer
	34 Whittington Terrace, Astonbury
Age:	40
Prescription:	Cod liver oil 30% v/v emulsion
Directions:	10 mL tds ac
Mitte:	200 mL

1. **Use of the product**
   Used as a source of vitamins A and D. Also contains several unsaturated fatty acids (*Martindale* 35th edn, p 1774).

2. **Is it safe and suitable for the intended purpose?**
    As a prophylactic against rickets, the dose of Cod Liver Oil is
    not more than 10 mL/day, allowance being made for vitamin
    D obtained from other sources (*British Pharmaceutical Codex*
    1973, p 124). The dose of this product on the prescription is
    10 mL three times a day, giving a total daily dose of 30 mL. As
    the product will contain 30% cod liver oil, this gives a dose of
    $(30 \div 100) \times 30 = 9$ mL cod liver oil per day. Therefore the dose
    is safe and suitable.
3. **Calculation of formula for preparation**
    Prepare 200 mL of Cod Liver Oil 30% emulsion.

Product formula

	Master	200 mL
Cod Liver Oil BP	30 mL	60 mL
Acacia BP	qs	qs
Double Strength Chloroform Water BP	50 mL	100 mL
Freshly boiled and cooled purified water	to 100 mL	to 200 mL

The quantity of emulsifying agent (Acacia BP) required to produce
200 mL of the emulsion must be calculated.

Formula for primary emulsion

Cod Liver Oil BP is a fixed oil. Therefore the primary emulsion
ratio is:

Oil	:	Water	:	Gum
4	:	2	:	1

60 mL of Cod Liver Oil BP is required, therefore 4 parts = 60 mL.
1 part will therefore be $60 \div 4 = 15$.
Therefore:
The amount of freshly boiled and cooled purified water needed
$= 2 \times 15$ mL $= 30$ mL.
The amount of Acacia BP required $= 15$ g.

Therefore the product formula for 200 mL of Cod Liver Oil 30%
emulsion is:

	200 mL
Cod Liver Oil BP	60 mL
Acacia BP	15 g
Double Strength Chloroform Water BP	100 mL
Freshly boiled and cooled purified water	to 200 mL

Interim formula for Double Strength Chloroform Water BP

Concentrated Chloroform Water BPC 1959	5 mL
Freshly boiled and cooled purified water	to 100 mL

4. **Method of preparation**
a. Solubility where applicable
    Not applicable.
b. Vehicle/diluent
    As emulsions are particularly susceptible to microbial contamination, Double Strength Chloroform Water BP will be used as the vehicle at a concentration of 50%. Freshly boiled and cooled purified water will be used as the remainder of the vehicle. As freshly boiled and cooled purified water is used in the product, it will also be used to make the Double Strength Chloroform Water BP.
c. Preservative
    Double Strength Chloroform Water BP is included in this product as the preservative.
d. Flavouring when appropriate
    No extra flavouring is required. In addition to preservative action, Double Strength Chloroform Water BP will give some flavouring.

The following method would be used to prepare 200 mL of Cod Liver Oil 30% emulsion from the formula above:

1. Calculate the composition of a convenient quantity of Double Strength Chloroform Water BP, sufficient to satisfy the formula requirements but also enabling simple, accurate measurement of the concentrated component.

Method of compounding for Double Strength Chloroform Water BP

a. In this case, 100 mL of Double Strength Chloroform Water BP is required. To prepare 100 mL Double Strength Chloroform Water BP, measure 5 mL of Concentrated Chloroform Water BPC 1959 accurately using a 5 mL conical measure.
b. Add approximately 90 mL of freshly boiled and cooled purified water to a 100 mL conical measure (i.e. sufficient water to enable dissolution of the concentrated chloroform component without reaching the final volume of the product).
c. Add the measured Concentrated Chloroform Water BPC 1959 to the water in the conical measure
d. Stir gently and then accurately make up to volume with freshly boiled and cooled purified water.

Tips

Although potable water would usually be used in the manufacture of Double Strength Chloroform Water BP, freshly boiled and cooled purified water is used here as emulsions are particularly susceptible to microbial contamination.

e. Visually check that no undissolved chloroform remains at the bottom of the measure.

2. Weigh 15 g of Acacia BP on a Class II or electronic balance.
3. Accurately measure 100 mL Double Strength Chloroform Water BP using a 100 mL measure.
4. Accurately measure 60 mL Cod Liver Oil BP in a conical measure.
5. Transfer the Cod Liver Oil BP to a clean dry porcelain mortar.
6. Measure 30 mL of Double Strength Chloroform Water BP (from the 100 mL measured in step 3).
7. Transfer the Acacia BP to the mortar and stir gently (approximately 3 stirs).
8. Add the 30 mL of Double Strength Chloroform Water BP to the mortar in one go.
9. Stir vigorously with the pestle in one direction only until the primary emulsion is formed.
10. Add more Double Strength Chloroform Water BP to the primary emulsion until the emulsion is pourable.
11. Transfer to an appropriate conical measure with rinsings.
12. Make up to volume with any remaining Double Strength Chloroform Water BP and freshly boiled and cooled purified water.
13. Stir and transfer to an amber flat medical bottle, label and dispense to the patient

5. **Choice of container**
    A plain amber bottle with a child-resistant closure would be most suitable as the preparation is an emulsion for internal use.
6. **Labelling considerations**
a. Title
    The product is unofficial, therefore the following title would be suitable: 'Cod liver oil 30% v/v emulsion'.

Tips

Remember the accurate preparation of the primary emulsion is crucial for the full emulsion to be satisfactorily produced.

Primary emulsion

Cod Liver Oil BP	60 mL
Double Strength Chloroform Water BP	30 mL
Acacia BP	15 g

Tips

Step 5: Ensure the measure is well drained as the quantities to be used are critical in the formation of the primary emulsion.

Step 7: This step is to wet the Acacia BP. Gentle stirring is required to ensure that there is no heat production that may denature the gum and prevent the formation of the emulsion.

Step 9: A characteristic clicking sound will be heard when the primary emulsion is formed. Remember the whiter the primary emulsion, the better it is formed.

Step 10: The Double Strength Chloroform Water BP needs to be added drop by drop to the primary emulsion until it is pourable to ensure that the primary emulsion does not crack.

Step 12: Even though there is a preservative in the preparation, freshly boiled and cooled purified water is used here as emulsions are particularly susceptible to microbial contamination.

b. Quantitative particulars
Quantitative particulars are required as the product is
unofficial. As this is a product for internal use, the quantitative
particulars will be expressed per dose (i.e. per 10 mL).

Each 10 mL contains:

Cod Liver Oil BP	3 mL
Acacia BP	0.75 g
Double Strength Chloroform Water BP	5 mL
Freshly boiled and cooled purified water	to 10 mL

c. Product-specific cautions (or additional labelling
requirements)
'Shake the bottle' will need to be added to the label as the
product is an emulsion and will need shaking before use to
ensure an accurate dose is measured.

d. Directions to patient – interpretation of Latin abbreviations
where necessary
'Take TWO 5 mL spoonfuls THREE times a day before food.'

e. Recommended *British National Formulary* cautions when
suitable
Not applicable.

f. Discard date.
The product is an emulsion and so will attract a 4-week
discard date.

g. Sample label (you can assume that the name and address of the
pharmacy and the words 'Keep out of the reach and sight of
children' are pre-printed on the label):

Cod liver oil 30% v/v emulsion   **200 mL**

Take TWO 5 mL spoonfuls THREE times a
day before food.

Shake the bottle

Do not use after (4 weeks)

Each 10 mL contains:

Cod Liver Oil BP	3 mL
Acacia BP	0.75 g
Double Strength Chloroform Water BP	5 mL
Freshly boiled and cooled purified water	to 10 mL

Mrs Fiona Archer   Date of dispensing

---

7. Advice to patient
The patient would be advised to take two 5 mL spoonfuls
three times a day before food. In addition, the discard date
and the need to shake the bottle would be highlighted to the
patient.

Example 4.2
The preparation of a magistral formulation from a doctor's prescription

You receive a prescription in your pharmacy with the following details:

Patient:	Mr James Taylor, 123 Station Road, Astonbury
Age:	23
Prescription:	Arachis Oil BP 40% emulsion with peppermint flavouring
Directions:	15 mL bd
Mitte:	150 mL

1. **Use of the product**
   Arachis Oil BP is nutritious, demulcent and mildly laxative (*British Pharmaceutical Codex* 1973, p 31).
2. **Is it safe and suitable for the intended purpose?**
   Calogen is an emulsion for disease-related malnutrition and malabsorption states (*British National Formulary* 61st edn, p 920) and contains arachis oil at a concentration of 50%. Therefore this formula is safe and suitable for use.
3. **Calculation of formula for preparation**
   Prepare 150 mL of Arachis Oil BP 40% emulsion with a peppermint flavouring.
       Before deciding the formula for the emulsion, the type and quantity of flavouring must be decided upon. Concentrated Peppermint Emulsion BP is a suitable flavouring and the dose is 0.25-1 mL. The dose of the emulsion is 15 mL bd which means each individual dose is 15 mL, therefore in 150 mL there would be 10 doses. The amount of Concentrated Peppermint Emulsion BP that would be suitable to use would be $10 \times 0.25 = 2.5$ mL.

Product formula

	Master	50 mL	150 mL
Arachis Oil BP	40 mL	20 mL	60 mL
Acacia BP	qs	qs	qs
Concentrated Peppermint Emulsion BP	qs	qs	2.5 mL
Double Strength Chloroform Water BP	50 mL	25 mL	75 mL
Freshly boiled and cooled purified water	to 100 mL	to 50 mL	to 150 mL

The quantity of emulsifying agent (Acacia BP) required to produce 150 mL of the emulsion must be calculated.

Formula for primary emulsion
Arachis oil is a fixed oil. Therefore the primary emulsion ratio is:

Oil :	Water :	Gum
4 :	2 :	1

60 mL of Arachis Oil BP is required, therefore 4 parts = 60 mL. 1 part will therefore be 60 ÷ 4 = 15.

Therefore:

The amount of freshly boiled and cooled purified water needed = 2 × 15 mL = 30 mL.

The amount of Acacia BP required = 15 g.

Therefore the product formula for 150 mL of Arachis Oil BP 40% emulsion is:

	150 mL
Arachis Oil BP	60 mL
Acacia BP	15 g
Concentrated Peppermint Emulsion BP	2.5 mL
Double Strength Chloroform Water BP	75 mL
Freshly boiled and cooled purified water	to 150 mL

Interim formula for Double Strength Chloroform Water BP:

Concentrated Chloroform Water BPC 1959	5 mL
Freshly boiled and cooled purified water	to 100 mL

4. **Method of preparation**
a. Solubility where applicable

   Not applicable as no solids will need to be dissolved during the preparation of this product.

b. Vehicle/diluent

   As emulsions are particularly susceptible to microbial contamination, Double Strength Chloroform Water BP will be used as the vehicle at a concentration of 50%. Freshly boiled and cooled purified water will be used as the remainder of the vehicle. As freshly boiled and cooled purified water is used in the product, it will also be used to make the Double Strength Chloroform Water BP.

c. Preservative

   Double Strength Chloroform Water BP is included in this product as the preservative.

d. Flavouring when appropriate

   The prescriber has requested that the emulsion is flavoured with peppermint flavouring. In this case, Concentrated Peppermint Emulsion BP is used as the source of the peppermint flavouring.

The following method would be used to prepare 150 mL of Arachis Oil BP 40% emulsion from the formula above:

1. Calculate the composition of a convenient quantity of Double Strength Chloroform Water BP, sufficient to satisfy

the formula requirements but also enabling simple, accurate measurement of the concentrated component.

Method of compounding for Double Strength Chloroform Water BP

a. In this case, 75 mL of Double Strength Chloroform Water BP is required and so it would be sensible to prepare 100 mL. To prepare 100 mL Double Strength Chloroform Water BP, measure 5 mL of Concentrated Chloroform Water BPC 1959 accurately using a 5 mL conical measure.

b. Add approximately 90 mL of freshly boiled and cooled purified water to a 100 mL conical measure (i.e. sufficient water to enable dissolution of the concentrated chloroform component without reaching the final volume of the product).

c. Add the measured Concentrated Chloroform Water BPC 1959 to the water in the conical measure.

d. Stir gently and then accurately make up to volume with freshly boiled and cooled purified water.

e. Visually check that no undissolved chloroform remains at the bottom of the measure.

2. Weigh 15 g of Acacia BP on a Class II or electronic balance.

3. Accurately measure 75 mL Double Strength Chloroform Water BP using a 100 mL measure.

4. Accurately measure 60 mL Arachis Oil BP in a conical measure.

5. Transfer the Arachis Oil BP to a clean dry porcelain mortar.

6. Measure 30 mL of Double Strength Chloroform Water BP (from the 75 mL measured in step 3).

7. Transfer the Acacia BP to the mortar and stir gently (approximately 3 stirs).

8. Add the 30 mL of Double Strength Chloroform Water BP to the mortar all in one go.

9. Stir vigorously with the pestle in one direction only until the primary emulsion is formed.

10. Add more Double Strength Chloroform Water BP to the primary emulsion until the emulsion is pourable.

11. Transfer to an appropriate conical measure with rinsings.

Tips

Remember the accurate preparation of the primary emulsion is crucial for the full emulsion to be satisfactorily produced.

Primary emulsion

Cod Liver Oil BP	60 mL
Double Strength Chloroform Water BP	30 mL
Acacia BP	15 g

Tips

The Concentrated Peppermint Emulsion BP is added just prior to making up to volume as it is a volatile ingredient. In addition, if the Concentrated Peppermint Emulsion BP is added any earlier, there is a risk that the emulsion may become unstable.

12. Measure 2.5 mL of Concentrated Peppermint Emulsion BP using an appropriate syringe.
13. Add the Concentrated Peppermint Emulsion BP to the emulsion in the conical measure.
14. Make up to volume with any remaining Double Strength Chloroform Water BP and freshly boiled and cooled purified water.
15. Stir and transfer to an amber flat medical bottle, label and dispense to the patient.

5. **Choice of container**
   A plain amber bottle with a child-resistant closure would be most suitable as the preparation is an emulsion for internal use.

6. **Labelling considerations**
   a. Title
      The product is unofficial, therefore the following title would be suitable: 'Arachis oil 40% v/v emulsion'.
   b. Quantitative particulars
      Quantitative particulars are required as the product is unofficial. As this is a product for internal use, the quantitative particulars will be expressed per dose (i.e. per 15 mL).

Each 15 mL contains:

Arachis Oil BP	6 mL
Acacia BP	1.5 g
Concentrated Peppermint Emulsion BP	0.25 mL
Double Strength Chloroform Water BP	7.5 mL
Freshly boiled and cooled purified water	to 15 mL

   c. Product-specific cautions (or additional labelling requirements)
      'Shake the bottle' will need to be added to the label as the product is an emulsion and will need shaking before use to ensure an accurate dose is measured.
   d. Directions to patient – interpretation of Latin abbreviations where necessary 'Take THREE 5 mL spoonfuls TWICE a day'.
   e. Recommended *British National Formulary* cautions when suitable
      Not applicable.
   f. Discard date
      The product is an emulsion and so will attract a 4-week discard date.
   g. Sample label (you can assume that the name and address of the pharmacy and the words 'Keep out of the reach and sight of children' are pre-printed on the label):

Arachis oil 40% v/v emulsion   **150 mL**

Take THREE 5 mL spoonfuls TWICE a day.

Shake the bottle

Do not use after (4 weeks)

Each 15 mL contains:

Arachis Oil BP	6 mL
Acacia BP	1.5 g
Concentrated Peppermint Emulsion BP	0.25 mL
Double Strength Chloroform Water BP	7.5 mL
Freshly boiled and cooled purified water	to 15 mL

Mr James Taylor   Date of dispensing

---

7. **Advice to patient**

   The patient would be advised to take three 5 mL spoonfuls twice a day. In addition, the discard date and the need to shake the bottle would be highlighted to the patient.

Example 4.3
The preparation of a magistral formulation from a doctor's prescription

You receive a prescription in your pharmacy with the following details:

Patient:	Ms Jayne Smith, 76 Oak Crescent, Astonbury
Age:	45
Prescription:	Liquid Paraffin 15% emulsion
Directions:	15 mL tds
Mitte:	200 mL

1. **Use of the product**

   Used as a laxative to treat constipation (*British National Formulary* 61st edn, p 71).

2. **Is it safe and suitable for the intended purpose?**

   The product is similar to Liquid Paraffin Emulsion BPC, which contains 5 mL of Liquid Paraffin BP per 10 mL, with a dose of 10–30 mL (*British Pharmaceutical Codex* 1973, pp 678–679). This would equate to 5–15 mL of Liquid Paraffin BP per dose (three to four times a day by convention).

   The prescription requests 15 mL of a 15% emulsion to be taken three times a day. This equates to $(15 \div 100) \times 15 = 2.25$ mL Liquid Paraffin BP three times a day. This dose is lower than the reference dose, but confirmation of the condition of the patient (for example, the patient may be elderly or infirm) and/or the condition being treated (for example, the product may be prescribed to treat mild constipation) will indicate whether the product is safe and suitable for the intended purpose.

3. **Calculation of formula for preparation**
   Prepare 200 mL of Liquid Paraffin BP 15% emulsion.

Product formula

	Master	200 mL
Liquid Paraffin BP	15 mL	30 mL
Acacia BP	qs	qs
Double Strength Chloroform Water BP	50 mL	100 mL
Freshly boiled and cooled purified water	to 100 mL	to 200 mL

The quantity of emulsifying agent (Acacia BP) required to produce 200 mL of the emulsion must be calculated.

Formula for primary emulsion

Liquid Paraffin BP is a mineral oil. Therefore the primary emulsion ratio is:

Oil :   Water :   Gum

3 :   2   : 1

30 mL of Liquid Paraffin BP is required, therefore 3 parts = 30 mL.

1 part will therefore be $30 \div 3 = 10$.

Therefore:

The amount of freshly boiled and cooled purified water needed = $2 \times 10 = 20$ mL.

The amount of Acacia BP required = 10 g.

Therefore the product formula for 200 mL of Liquid Paraffin BP 30% emulsion is:

	200 mL
Liquid Paraffin BP	30 mL
Acacia BP	10 g
Double Strength Chloroform Water BP	100 mL
Freshly boiled and cooled purified water	to 200 mL

Interim formula for Double Strength Chloroform Water

Concentrated Chloroform Water BPC	1959 5 mL
Freshly boiled and cooled purified water	to 100 mL

4. **Method of preparation**
a. **Solubility where applicable**
   Not applicable.
b. **Vehicle/diluent**
   As emulsions are particularly susceptible to microbial contamination, Double Strength Chloroform Water BP will be used as the vehicle at a concentration of 50%. Freshly boiled

and cooled purified water will be used as the remainder of the vehicle. As freshly boiled and cooled purified water is used in the product, it will also be used to make the Double Strength Chloroform Water BP.

c. **Preservative**
Double Strength Chloroform Water BP is included in this product as the preservative.

d. **Flavouring when appropriate**
No extra flavouring is required. In addition to preservative action Double Strength Chloroform Water BP will give some flavouring.

The following method would be used to prepare 200 mL of Liquid Paraffin BP 15% emulsion from the formula above:

1. Calculate the composition of a convenient quantity of Double Strength Chloroform Water BP, sufficient to satisfy the formula requirements but also enabling simple, accurate measurement of the concentrated component.

Method of compounding for Double Strength Chloroform Water BP

a. In this case, 100 mL of Double Strength Chloroform Water BP is required. To prepare 100 mL Double Strength Chloroform Water BP, measure 5 mL of Concentrated Chloroform Water BPC 1959 accurately using a 5 mL conical measure.

b. Add approximately 90 mL of freshly boiled and cooled purified water to a 100 mL conical measure (i.e. sufficient water to enable dissolution of the concentrated chloroform component without reaching the final volume of the product).

c. Add the measured Concentrated Chloroform Water BPC 1959 to the water in the conical measure.

d. Stir gently and then accurately make up to volume with freshly boiled and cooled purified water.

e. Visually check that no undissolved chloroform remains at the bottom of the measure.

2. Weigh 10 g of Acacia BP on a Class II or electronic balance.

3. Accurately measure 100 mL Double Strength Chloroform Water BP using a 100 mL measure.

Tips

Remember the accurate preparation of the primary emulsion is crucial for the full emulsion to be satisfactorily produced.

Primary emulsion

Liquid Paraffin BP Double Strength Chloroform Water BP	30 mL
	20 mL
Acacia BP	10 g

4. Accurately measure 30 mL Liquid Paraffin BP in a conical measure.
5. Transfer the Liquid Paraffin BP to a clean dry porcelain mortar.
6. Measure 20 mL of Double Strength Chloroform Water BP (from the 100 mL measured in step 3).
7. Transfer the Acacia BP to the mortar and stir gently (approximately 3 stirs).
8. Add the 20 mL of Double Strength Chloroform Water BP to the mortar all in one go.
9. Stir vigorously with the pestle in one direction only until the primary emulsion is formed.
10. Add more Double Strength Chloroform Water BP to the primary emulsion until the emulsion is pourable.
11. Transfer to an appropriate conical measure with rinsings.
12. Make up to volume with any remaining Double Strength Chloroform Water BP and freshly boiled and cooled purified water.
13. Stir and transfer to an amber flat medical bottle, label and dispense to the patient.

5. **Choice of container**

A plain amber bottle with a child-resistant closure would be most suitable as the preparation is an emulsion for internal use.

6. **Labelling considerations**

a. Title

The product is unofficial, therefore the following title would be suitable: 'Liquid paraffin 15% v/v emulsion'.

b. **Quantitative particulars**

Quantitative particulars are required as the product is unofficial. As this is a product for internal use, the quantitative particulars will be expressed per dose (i.e. per 15 mL).

Each 15 mL contains:

Liquid Paraffin BP	2.25 mL
Acacia BP	0.75 g
Double Strength Chloroform Water BP	7.5 mL
Freshly boiled and cooled purified water	to 15 mL

c. Product-specific cautions (or additional labelling requirements) 'Shake the bottle' will need to be added to the label as the product is an emulsion and will need shaking before use to ensure an accurate dose is measured.

d. Directions to patient – interpretation of Latin abbreviations where necessary

'Take THREE 5 mL spoonfuls THREE times a day'.

e. Recommended *British National Formulary* cautions when suitable

Not applicable.

f. Discard date
    This product is an emulsion and so will attract a 4-week
    discard date.
g. Sample label (you can assume that the name and address of the
    pharmacy and the words 'Keep out of the reach and sight of
    children' are preprinted on the label):

Liquid paraffin 15% v/v emulsion   **200 mL**

Take THREE 5 mL spoonfuls THREE times a day.

Do not use after (4 weeks)

Shake the bottle

Each 15 mL contains:

Liquid Paraffin BP	2.25 mL
Acacia BP	0.75 g
Double Strength Chloroform Water BP	7.5 mL
Freshly boiled and cooled purified water	to 15 mL

Ms Jayne Smith   Date of dispensing

---

7. **Advice to patient**
    The patient would be advised to take three 5 mL spoonfuls
    three times a day. In addition, the discard date and the need to
    shake the bottle would be highlighted to the patient.

Example 4.4
The preparation of Liquid Paraffin Emulsion BP

You receive a prescription in your pharmacy with the following
details:

Patient:	Mrs Daisy Marlow, 72 Oak Street, Astonbury
Age:	50
Prescription:	Liquid Paraffin Emulsion BP
Directions:	10 mL tds
Mitte:	100 mL

1. **Use of the product**
    Used as a laxative to treat constipation (*British National
    Formulary* 61st edn, p 71).
2. **Is it safe and suitable for the intended purpose?**
    The product is the same as Liquid Paraffin Emulsion BPC,
    which contains 5 mL of Liquid Paraffin BP per 10 mL, with a
    dose of 10–30 mL (*British Pharmaceutical Codex* 1973, pp 678–
    679). Therefore, the product and dose are suitable and suitable
    for the intended purpose.
3. **Calculation of formula for preparation**
    Prepare 100 mL of Liquid Paraffin Emulsion BP.

Product formula (from the British Pharmacopoeia 2007, p 2818)

	Master	100 mL
Liquid Paraffin BP	500 mL	50 mL
Vanillin BP	500 mg	50 mg
Chloroform BP	2.5 mL	0.25 mL
Benzoic Acid Solution BP	20 mL	2 mL
Methylcellulose 20 BP	20 g	2 g
Saccharin Sodium BP	50 mg	5 mg
Freshly boiled and cooled purified water	to 1000 mL	to 100 mL

4. Method of preparation

a. Solubility where applicable

As the method for preparation is given in the *British Pharmacopoeia* for this product, it is not necessary to check each individual solubility.

b. Vehicle/diluent

As emulsions are particularly susceptible to microbial contamination, freshly boiled and cooled purified water will be used as per the product formula.

c. Preservative

Chloroform BP and Benzoic Acid Solution BP are included in this product as preservatives.

d. Flavouring when appropriate

No extra flavouring is required. In addition to preservative action Chloroform Water BP will give some flavouring. Vanillin BP and Saccharin BP will also flavour the product.

The following method would be used to prepare 100 mL of Liquid Paraffin Emulsion BP from the formula above (from the *British Pharmacopoeia* 2007, p 2818):

1. Heat about 12 mL of freshly boiled and cooled purified water.
2. Weigh 2 g Methylcellulose 20 BP on a Class II or electronic balance.
3. Add the Methylcellulose 20 BP to the heated water.
4. Allow to stand for about 30 minutes to hydrate.
5. Add sufficient freshly boiled and cooled purified water in the form of ice to produce 35 mL and stir.
6. Measure 2 mL of Benzoic Acid Solution BP using a syringe.
7. Measure 0.25 mL of Chloroform BP.
8. Weigh 50 mg Vanillin BP using a sensitive electronic balance
9. Mix the Chloroform BP and Benzoic Acid Solution BP together.

10. Dissolve the Vanillin BP in the benzoic acid and chloroform mixture.

11. Add this solution to the previously prepared methylcellulose mucilage and stir for 5 minutes.

12. Prepare a sodium saccharin trituration and add sufficient (5 mL) to provide 5 mg of Sodium Saccharin BP to the mixture.

13. Make the volume of the mucilage mixture up to 50 mL with freshly boiled and cooled purified water.

14. Measure 50 mL of Liquid Paraffin BP in a 50 mL conical measure.

15. Mix the 50 mL mucilage and 50 mL of Liquid Paraffin BP together and stir constantly.

16. Pass through a homogeniser to make the emulsion more stable.

17. Transfer to an amber flat medical bottle with a child-resistant closure, label and dispense.

Figure 4.2. A homogeniser.

Tips

Vanillin BP is only slightly soluble in water but freely soluble in alcohol and soluble in ether. It is therefore more soluble in organic solvents so is added to the chloroform-containing mixture.

Tips

The amount of Sodium Saccharin BP cannot be accurately weighed, therefore a trituration must be prepared. Water is the diluent chosen as this is also the vehicle for the emulsion.

Trituration for Sodium Saccharin
Saccharin Sodium BP   150 mg
Freshly boiled and   to 150 mL
cooled purified water

Therefore 5 mL of the trituration will contain 5 mg of Sodium Saccharin BP.

Tips

The stability of an emulsion is increased with smaller globule size of the disperse phase. When an emulsion is passed through a homogeniser (Figure 4.2) the emulsion is forced through a fine opening to apply shearing forces to reduce the size of the globules. Although many extemporaneously prepared emulsions may not require the use of a homogeniser, this step may aid in retarding or preventing creaming of the emulsion on long standing.

5.   Choice of container

A plain amber bottle with a child-resistant closure would be most suitable as the preparation is an emulsion for internal use.

6. **Labelling considerations**

a. Title

The product is official, therefore the following title would be suitable: 'Liquid Paraffin Emulsion BP'.

b. **Quantitative particulars**

Quantitative particulars are not required as the product is official.

c. **Product-specific cautions (or additional labelling requirements)**

'Shake the bottle' will need to be added to the label as the product is an emulsion and will need shaking before use to ensure an accurate dose is measured.

d. **Directions to patient – interpretation of Latin abbreviations where necessary**

'Take TWO 5 mL spoonfuls THREE times a day'.

e. **Recommended *British National Formulary* cautions when suitable**

Not applicable.

f. **Discard date**

The product is an emulsion and so will attract a 4-week discard date.

g. **Sample label (you can assume that the name and address of the pharmacy and the words 'Keep out of the reach and sight of children' are pre-printed on the label):**

Liquid Paraffin Emulsion BP	**100 mL**
Take TWO 5 mL spoonfuls THREE times a day.	
Shake the bottle	
Do not use after (4 weeks)	
Mrs Daisy Marlow	Date of dispensing

7. **Advice to patient**

The patient would be advised to take two 5 mL spoonfuls three times a day. In addition, the discard date and the need to shake the bottle would be highlighted.

Self-assessment

1. What is the definition of an emulsion?

2. Why is it important to include a preservative in an oral emulsion?

3. Why is freshly boiled and cooled purified water used when preparing oral emulsions?

4. Fill in the following table:

Proportion in primary emulsion

Type of oil	Oil	Aqueous	Gum
Fixed oil			
Volatile oil			
Mineral oil			

5. Which of the following is a mineral oil?
a. Peppermint Oil BP
b. Corn Oil BP
c. Arachis Oil BP
d. Liquid Paraffin BP

6. Which of the following cannot be used as an emulgent?
a. Acacia BP
b. Chlorocresol BP
c. Tragacanth BP
d. Methylcellulose 20 BP

7. Which of the following is a volatile oil?
a. Almond Oil BP
b. Liquid Paraffin BP
c. Peppermint Oil BP
d. Maize Oil BP

8. What would be the proportions of oil:aqueous:emulsifier if making a primary emulsion containing a fixed oil?
a. 4:2:1
b. 3:2:1
c. 2:2:1
d. 1:2:1

9. Which of the following statements is correct?
a. A creamed emulsion will reform on shaking.
b. A creamed emulsion will not reform on shaking.
c. A cracked emulsion will reform on shaking.
d. A phase inverted emulsion will revert to its original form on shaking.

10. You are asked to prepare 100 mL of emulsion containing Liquid Paraffin BP 36%, Amaranth BP 0.5% and Syrup BP 10%. How much Acacia BP would be needed to prepare the primary emulsion?
a. 9 g
b. 10 g
c. 12 g
d. 18 g

11. How much Double Strength Chloroform Water BP would be included in the master formula for this product (from Question 10)?
a. None
b. 24 mL
c. 36 mL
d. 50 mL

12. How much Acacia BP would be needed to prepare 50 mL of an emulsion containing 20% Peppermint Oil BP?
a. 2.5 g
b. 5 g
c. 10 g
d. 20 g

13. A suitable discard date for an emulsion is:
a. 1 week
b. 2 weeks
c. 4 weeks
d. 3 months

14. What specific instruction should be included on the label of all emulsions?
a. 'For external use only.'
b. 'Not to be taken.'
c. 'Shake the bottle.'
d. 'Store in a cool place.'

Formulation questions

This section contains details of extemporaneous products to be made in the same way as the examples earlier in this chapter. For each example, provide answers using the following sections:

1. **Use of the product**
2. **Is it safe and suitable for the intended purpose?**
3. **Calculation of formula for preparation**
4. **Method of preparation**
a. Solubility where applicable
b. Vehicle/diluent
c. Preservative
d. Flavouring when appropriate
5. **Choice of container**
6. **Labelling considerations**
a. Title
b. Quantitative particulars
c. Product-specific cautions (or additional labelling requirements)
d. Directions to patient – interpretation of Latin abbreviations where necessary
e. Recommended *British National Formulary* cautions when suitable

f. Discard date
g. Sample label (you can assume that the name and address of the pharmacy and the words 'Keep out of the reach and sight of children' are pre-printed on the label)
7. **Advice to patient**

15. **You receive a prescription in your pharmacy with the following details:**

Patient:	Miss Sophie Jones, 29 Star Terrace, Astonbury
Age:	28
Prescription:	Maize oil 30% v/v emulsion
Directions:	15 mL tds
Mitte:	100 mL

16. **You receive a prescription in your pharmacy with the following details:**

Patient:	Ms Petra Williams, 526 High Street, Astonbury
Age:	45
Prescription:	Castor Oil 30% v/v emulsion
Directions:	20 mL nocte
Mitte:	50 mL

chapter 5
Creams

Overview

Upon completion of this chapter, you should be able to:

- prepare a cream from first principles
- incorporate solids and liquids into a cream base
- select an appropriate container in which to package a cream
- prepare an appropriate label for a cream.

Introduction and overview of creams

In pharmacy the term 'cream' is reserved for external preparations. Creams are viscous semi-solid emulsions for external use. Medicaments can be dissolved or suspended in creams.

A cream may be 'water-in-oil' or 'oil-in-water' depending on the emulsifying agent used. A cream is always miscible with its continuous phase.

British Pharmacopoeia (BP) definition

Creams are formulated to provide preparations that are essentially miscible with the skin secretion. They are intended to be applied to the skin or certain mucous membranes for protective, therapeutic or prophylactic purposes, especially where an occlusive effect is not necessary.

General method

Terminology used in the preparation of creams, ointments, pastes and gels

The following are common terms that are used within the extemporaneous preparation of creams and in the extemporaneous preparation of ointments, pastes and gels (see Chapter 6).

Trituration

This is the term applied to the incorporation, into the base, of finely divided insoluble powders or liquids. The powders are placed on the tile and the base is incorporated using the 'doubling-up' technique. Liquids are usually incorporated by placing a small amount of ointment base on a tile

Definition

Water-in-oil creams (oily creams) as bases
These are produced by emulsifying agents of natural origin, e.g. beeswax, wool alcohols or wool fat. These bases have good emollient properties. They are creamy, white or translucent and rather stiff.

Oil-in-water creams (aqueous creams) as bases
These are produced by synthetic waxes, e.g. macrogol and cetomacrogol. They are the best bases to use for rapid absorption and penetration of drugs. They are thin, white and smooth in consistency.

and making a 'well' in the centre. Small quantities of liquid are then added and mixed in. Take care not to form air pockets that contain liquid, which if squeezed when using an inappropriate mixing action will spray fluid on the compounder and surrounding area.

Trituration can be successfully achieved using a mortar but this method is usually reserved for large quantities.

Levigation

This is the term applied to the incorporation into the base of insoluble coarse powders. It is often termed 'wet grinding'. It is the process where the powder is rubbed down with either the molten base or semi-solid base. A considerable shearing force is applied to avoid a gritty product.

The preparation of a cream from first principles

1. As with other types of emulsion, hygiene is extremely important and all surfaces, spatulas and other equipment must be thoroughly cleaned with industrial denatured alcohol (IDA). IDA is better than freshly boiled and cooled purified water as it will quickly evaporate, leaving no residue.
2. Always make an excess as it is never possible to transfer the entire cream into the final container.
3. Determine which of the ingredients are soluble in/miscible with the aqueous phase and which with the oily phase. Dissolve the water-soluble ingredients in the aqueous phase.
4. Melt the fatty bases in an evaporating dish over a water bath (Figure 5.1) at the lowest possible temperature. Start with the base with the highest melting point. These should then be cooled to 60°C (overheating can denature the emulsifying agent and the stability of the product can be lost).
5. Substances that are soluble/miscible with the oily phase should then be stirred into the melt.
6. The temperature of the aqueous phase should then be adjusted to 60°C.
7. The disperse phase should then be added to the continuous phase at the same temperature.
8. Stir the resulting emulsion vigorously without incorporating air, until the product sets. Do not hasten cooling as this produces a poor product.

See Creams video for a demonstration of the preparation of a cream from first principles.

The incorporation of ingredients into a cream base

In addition to the preparation of a cream from first principles, it is common to incorporate either liquid or solid ingredients into a cream base.

Figure 5.1.
A water bath and heater.

The incorporation of solids into a cream base

If the cream base has been prepared from first principles
(see above), the solid can be incorporated into the cream as it cools.
Alternatively, if using a pre-prepared base, soluble and insoluble
solids may be incorporated using the method employed for insoluble
solids.

- **Soluble solids** should be added to the molten cream at
  the lowest possible temperature and the mixture stirred until
  cold.
- **Insoluble solids** should be incorporated using a glass tile and
  spatula (Figure 5.2). If there is more than one powder to be
  added, these should be triturated together in a mortar using the
  'doubling-up' technique prior to transfer to a glass tile.
- **Coarse powders.** A minimum quantity of cream should be
  placed in the centre of the glass tile and used to levigate the
  powders. A considerable lateral shearing force should be
  applied to avoid a gritty product

  The powder/fatty base mixture may then either be returned
  to the evaporating basin with the remaining cream and stirred
  until cold or the remaining cream in the evaporating basin
  may be allowed to cool and triturated with the powder/cream
  mixture on the tile.

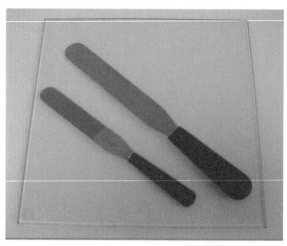

Figure 5.2.
An ointment tile and spatulas.

– **Fine powders** may be triturated into the otherwise finished cream on a glass tile. Small amounts of powder should be added to an equal amount of cream, i.e. using the 'doubling-up' technique. These should be well triturated.

 See Ointments video for a demonstration of the incorporation of a powder into a base.

The incorporation of liquids into a cream base

■ **Non-volatile, miscible liquids** may be mixed with the molten cream in the evaporating basin. Alternatively, if a pre-prepared base is used, then incorporate as for volatile or immiscible liquids.

■ **Volatile or immiscible liquids**, e.g. coal tar solutions, should be triturated with the cream on the glass tile.

A very small amount of the cream should be placed on the glass tile and a 'well' made in the centre. Traditionally, small quantities of liquid should be gently folded in to avoid splashing. An alternative method is to spread a small amount of the cream on the tile and then score it with a spatula. Then add small quantities of the liquid and fold into the base gently.

If using coal tar or other volatile ingredients, these should not be weighed until immediately before use and the beaker in which it has been weighed should be covered with a watch glass to prevent evaporation. In addition, always remember that volatile ingredients should not be added to molten bases.

See Ointments video for a demonstration of the incorporation of a liquid into a base.

Worked examples

Example 5.1

The preparation of Cetrimide Cream BP

You receive a prescription in your pharmacy with the following details:

Patient:	Mrs Samantha Fuller, 4 Park Street, Astonbury
Age:	42
Prescription:	Cetrimide Cream BP
Directions:	Apply to the abrasion tds
Mitte:	20 g

1. **Use of the product**
   Aqueous solutions and creams containing 0.1–1% cetrimide are used for the treatment of wounds and burns, for preoperative cleansing of the skin and for the removal of scabs and crusts in skin disease (*British Pharmaceutical Codex* 1973, p 88).

2. **Is it safe and suitable for the intended purpose?**
   This is an official preparation, therefore the formula is safe and suitable for purpose.

3. **Calculation of formula for preparation**
   Prepare 20 g Cetrimide Cream BP.

Product formula
(from the *British Pharmacopoeia* 2007, p 2406)

	Master	100 g	10 g	30 g
Cetrimide BP	5 g	500 mg	50 mg	150 mg
Cetostearyl Alcohol BP	50 g	5 g	500 mg	1.5 g
Liquid Paraffin BP	500 g	50 g	5 g	15 g
Freshly boiled and cooled purified water	445 g	44.5 g	4.45 g	13.35 g

4. **Method of preparation**
   a. Solubility where applicable
      Cetostearyl Alcohol BP is practically insoluble in water and, when melted, miscible with liquid paraffin (*Martindale* 35th edn, p 1846). Cetrimide BP is soluble in 2 parts water (*British Pharmacopoeia* 1988, p 111) and Liquid Paraffin BP is practically insoluble in water, sparingly soluble in ethanol and soluble with hydrocarbons (*British Pharmacopoeia* 1988, p 415).

Tips

The exact quantity cannot be prepared for a cream as losses will be encountered on transfer. Therefore, a suitable overage must be produced in order to dispense the required final amount.

Note that all liquid ingredients (including the Liquid Paraffin BP) must be weighed.

b. Vehicle/diluent

As creams are susceptible to microbial contamination, freshly boiled and cooled purified water is used as the vehicle.

c. Preservative

No preservative is included in this product (as per the product formula) and so freshly boiled and cooled purified water is used as the vehicle.

d. Flavouring when appropriate

Creams are for external use and so no flavouring is required.

The following method would be used to prepare 30 g of Cetrimide Cream BP from the formula above.

Note that the melting point of Cetostearyl Alcohol BP is 49–56°C (*British Pharmacopoeia* 1988, p 111):

1. Weigh 1.5 g Cetostearyl Alcohol BP on a Class II or electronic balance.
2. Weigh 15 g of Liquid Paraffin BP on a Class II or electronic balance.
3. Weigh 13.35 g of freshly boiled and cooled purified water on a Class II or electronic balance.
4. Weigh 150 mg of Cetrimide BP on a Class II or electronic balance.
5. Melt the Cetostearyl Alcohol BP in an evaporating basin over a water bath to a temperature no higher than 60°C.
6. Add the Liquid Paraffin BP to the molten Cetostearyl Alcohol BP and remove from the heat.
7. Stir to form the *oily* phase.
8. Transfer the freshly boiled and cooled purified water to a beaker and heat to 60°C.
9. Add the Cetrimide BP to the freshly boiled and cooled purified water and remove from the heat.
10. Stir to form the *aqueous* phase.
11. When the oily phase and the aqueous phase are both at about 60°C, add the aqueous phase to the oily phase with constant, not too vigorous stirring.
12. Stir until cool enough to pack.
13. Weigh 20 g of the product and pack into a collapsible tube or amber glass jar. Label and dispense.

5. Choice of container

A collapsible tube or plain amber jar would be most suitable.

Tips

Stirring is constant and not too vigorous to ensure that there are no 'cold spots' within the cream as these would hasten cooling in discrete areas and result in a lumpy cream.

6. Labelling considerations

a. Title

The product is official, therefore the following title would be suitable: 'Cetrimide Cream BP'.

b. Quantitative particulars

Quantitative particulars are not required as the product is official.

c. Product-specific cautions (or additional labelling requirements)
'For external use only' will need to be added to the label as the product is a cream for external use.

d. Directions to patient – interpretation of Latin abbreviations where necessary
'Apply to the abrasion THREE times a day'.

e. Recommended *British National Formulary* cautions when suitable
Not applicable.

f. Discard date
The product is a cream and so will attract a 4-week discard date.

g. Sample label (you can assume that the name and address of the pharmacy and the words 'Keep out of the reach and sight of children' are pre-printed on the label):

Cetrimide Cream BP	20 g
Apply to the abrasion THREE times a day.	
For external use only	
Do not use after (4 weeks)	
Mrs Samantha Fuller	Date of dispensing

7. **Advice to patient**
The patient would be advised to apply the cream to the abrasion three times a day. In addition, the discard date and the fact that the product is for external use only would be highlighted to the patient.

Example 5.2
The preparation of Salicylic Acid and Sulphur Cream BP
You receive a prescription in your pharmacy with the following details:

Patient:	Mr Ashok Patel, 14 Old Cross Road, Astonbury
Age:	34
Prescription:	Salicylic Acid and Sulphur Cream BP
Directions:	Apply mdu
Mitte:	20 g

1. **Use of the product**
Used to treat acne (*Martindale* 35th edn, pp 1451 and 1452).

2. **Is it safe and suitable for the intended purpose?**
This is an official preparation, therefore the formula is safe and suitable for purpose.

3. **Calculation of formula for preparation**
Prepare 20 g of Salicylic Acid and Sulphur Cream BP.

Product formula
(from the *British Pharmacopoeia* 1980, p 548):

	Master	100 g	10 g	30 g
Salicylic Acid BP	20 g	2 g	200 mg	600 mg
Precipitated Sulphur BP	20 g	2 g	200 mg	600 mg
Aqueous Cream BP	960 g	96 g	9.6 g	28.8 g

4. Method of preparation

a. Solubility where applicable

Not applicable.

b. Vehicle/diluent

Aqueous Cream BP is used as the base in this preparation as per the product formula.

c. Preservative

There is no preservative included as per the product formula.

d. Flavouring when appropriate

e. Creams are for external use and so no flavouring is required.

The following method would be used to prepare 30 g of Salicylic Acid and Sulphur Cream BP from the formula above:

1. Weigh 600 mg Salicylic Acid BP on a Class II or electronic balance.
2. Transfer to a glass mortar and grind with a pestle.
3. Weigh 600 mg Precipitated Sulphur BP on a Class II or electronic balance.
4. Add to the Salicylic Acid BP in the glass mortar and continue mixing with a pestle until a smooth well-mixed powder is formed.
5. Transfer the powder to a glass tile.
6. Weigh 28.8 g Aqueous Cream BP on a Class II or electronic balance.
7. Transfer the Aqueous Cream BP to the glass tile and triturate with the powders to produce a smooth product.
8. Weigh 20 g of the product and pack into a collapsible tube or amber glass jar. Label and dispense.

5. Choice of container

A collapsible tube or plain amber jar would be most suitable.

6. Labelling considerations

a. Title

The product is official, therefore the following title would be suitable: 'Salicylic Acid and Sulphur Cream BP'.

b. Quantitative particulars
    Quantitative particulars are not required as the product is official.
c. Product-specific cautions (or additional labelling requirements)
    'For external use only' will need to be added to the label as the
    product is a cream for external use.
d. Directions to patient – interpretation of Latin abbreviations
    where necessary
    'Apply as directed.'
e. Recommended *British National Formulary* cautions when
    suitable
    Not applicable.
f. Discard date
    The product is a cream and so will attract a 4-week discard date.
g. Sample label (you can assume that the name and address of the
    pharmacy and the words 'Keep out of the reach and sight of
    children' are pre-printed on the label):

Salicylic Acid and Sulphur Cream BP	**20 g**
Apply as directed.	
For external use only	
Do not use after (4 weeks)	
Mr Ashok Patel	Date of dispensing

7. **Advice to patient**
    The patient would be advised to apply the cream as directed.
    In addition, the discard date and the fact that the product
    is for external use only would be highlighted to the patient.
    If further direction for use is requested, the patient could
    be advised to apply the cream once or twice a day. Sulphur
    BP is usually applied once or twice a day and, although
    the concentration of Salicylic Acid BP is quite low when
    compared with preparations used in the treatment of warts
    and hard skin, it will act as a keratolytic and so a maximum
    application frequency of once or twice a day would be
    appropriate.

Example 5.3
The preparation of Dermovate Cream 25%

You receive a prescription in your pharmacy with the following
details:

Patient:	Mr Peter Johnson, 14 Vauxhall Parade, Astonbury
Age:	63
Prescription:	Dermovate Cream 1:3
Directions:	Apply bd
Mitte:	50 g

1. **Use of the product**
   Dermovate contains 0.05% clobetasol propionate, a corticosteroid for topical application for the short-term treatment of severe resistant inflammatory skin disorders. It is usually applied thinly 1–2 times daily for up to 4 weeks. Maximum 50 g of 0.05% preparation per week (*British National Formulary* 61st edn, p 711).
2. **Is it safe and suitable for the intended purpose?**
   It is reasonably common to dilute proprietary creams to produce less potent products for the treatment of inflammatory skin disorders. This is a 1 in 4 dilution (1:3) of a potent topical steroid. Therefore, so long as the dilution is stable (see below), the product will be safe and suitable for the intended purpose.
3. **Calculation of formula for preparation**
   Prepare 50 g of Dermovate Cream 1:3.

Product formula (i.e. the diluent to use) must be decided by the compounder. Refer to a diluent directory which in this case states that, although the dilutionof the product is not recommended by the manufacturer, in cases where it is insisted upon, the following may be used:

- Aqueous Cream BP
- Only stable if less than 50% of the resultant cream. Therefore unsuitable for this formulation.
- Buffered Cream BP
- May be used but can raise the pH of the resulting cream.
- Cetomacrogol Cream (Formula A) BPC
- No problems with dilution recorded.

Therefore the diluent of choice would be Cetomacrogol Cream (Formula A) BPC in this instance.

Note: Suitable sources to provide information on the dilution of creams and ointments would be:

- National Pharmacy Association *Diluent Directory*
- product data sheet (Summary of Product Characteristics – SPC)
- reports in the pharmaceutical literature
- personal contact with product manufacturer.

Product formula

	Master	100 g	10 g	60 g
Dermovate Cream	25%	25 g	2.5 g	15 g
Cetomacrogol Cream (Formula A) BPC	75%	75 g	7.5 g	45 g

4. **Method of preparation**
a. Solubility where applicable
   Not applicable.

b. Vehicle/diluent

Cetomacrogol Cream (Formula A) BPC is used as the diluent in this preparation.

c. Preservative

The product contains no additional preservative (apart from any preservative already present within the Dermovate Cream and pre-prepared Cetomacrogol Cream (Formula A) BPC).

d. Flavouring when appropriate

Creams are for external use and so no flavouring is required.

The following method would be used to prepare 60 g of Dermovate Cream 25% from the formula above:

1. Weigh 15 g Dermovate Cream on Class II or electronic balance.
2. Transfer to a glass tile.
3. Weigh 45 g Cetomacrogol Cream (Formula A) BPC on a Class II or electronic balance.
4. Transfer to the glass tile.
5. Triturate the Dermovate Cream with the Cetomacrogol Cream (Formula A) BPC using a spatula.
6. Weigh 50 g of the final cream on a Class II or electronic balance.
7. Pack into a collapsible tube or amber glass jar, label and dispense to the patient.

Tips

When triturating the creams together remember the principle of 'doubling-up' in order to achieve an adequate mix of the active Dermovate Cream and the base cream.

5. Choice of container

A collapsible tube or plain amber jar would be most suitable.

6. Labelling considerations

a. Title

The product is unofficial, therefore the following title would be suitable: 'Dermovate Cream 25% w/w'.

b. Quantitative particulars

Quantitative particulars are required as the product is unofficial.

They would be expressed per container (i.e. per 50 g):

The product contains:

Dermovate Cream	25%
Cetomacrogol Cream (Formula A) BPC	75%

c. Product-specific cautions (or additional labelling requirements)

'For external use only' will need to be added to the label as the product is a cream for external use.

d. Directions to patient – interpretation of Latin abbreviations where necessary

'Apply TWICE a day.'

e. Recommended *British National Formulary* cautions when
    suitable
    The *British National Formulary* (61st edn, p 711) recommends
    the following caution:
    Label 28 – 'Spread thinly on the affected skin only.'
f. Discard date
    The product is a cream and so would normally attract a 4-week
    discard date. However, as the product is a diluted proprietary
    cream, it is common to assign a shorter 2-week discard date.
g. Sample label (you can assume that the name and address of the
    pharmacy and the words 'Keep out of the reach and sight of
    children' are pre-printed on the label):

Dermovate Cream 25% w/w   **50 g**

Apply TWICE a day.
Spread thinly on the affected skin only
For external use only
Do not use after (2 weeks)

The product contains:

Dermovate Cream   25%
Cetomacrogol Cream (Formula A) BPC   75%
Mr Peter Johnson   Date of dispensing

---

7. **Advice to patient**
    The patient would be advised to apply the cream thinly/
    sparingly twice a day. In addition, the discard date and the fact
    that the product is for external use only would be highlighted
    to the patient.

Self-assessment

1. **Which statement is true?**
a. Creams are occlusive.
b. Creams are always water-in-oil emulsions.
c. The surfactant in creams allows better skin penetration than in ointments.
d. Creams are less susceptible than ointments to microbial contamination.

2. **How many grams of Betnovate Cream are contained in 200 g of a 1 in 20
    dilution of Betnovate Cream in Aqueous Cream BP?**
a. 2 g
b. 10 g
c. 20 g
d. 50 g

3. How many grams of Eumovate Cream are contained in 200 g of a 1 in 4 dilution of Eumovate Cream in Aqueous Cream BP?
a. 20 g
b. 25 g
c. 40 g
d. 50 g

4. How many grams of Betnovate Cream are contained in 300 g of a 1 in 5 dilution of Betnovate Cream in Cetomacrogol A Cream BPC?
a. 5 g
b. 6 g
c. 50 g
d. 60 g

5. How many grams of Dermovate Cream are contained in 200 g of a 1 in 10 dilution of Dermovate Cream in Aqueous Cream BP?
a. 10 g
b. 18 g
c. 20 g
d. 100 g

6. You receive a prescription for the following cream:

Calamine BP	4%
Zinc Oxide BP	3%
Emulsifying Wax BP	6%
Arachis Oil BP	30%
Water	to 100%

You are asked to prepare 50 g of the cream (the 50 g includes an overage). How much Arachis Oil BP must be used in this preparation?
e. 15 mL
f. 15 g
g. 30 mL
h. 30 g

7. What type of water would you use for preparing the above product (in Question 6)?
a. Double Strength Chloroform Water BP
b. Potable water
c. Freshly boiled and cooled purified water
d. Purified water

8. What would be the best way of expressing the quantitative particulars on a label for this product (from Question 6)?
a. As percentages
b. As quantities per 50 g
c. As quantities per 100 g
d. As a mixture of percentages and quantities

9. **The directions for use of the product in Question 6 are 'Applic ad lib ud'. On the label this would read as:**
a. 'Apply liberally.'
b. 'Apply as directed.'
c. 'Use as directed.'
d. 'Apply liberally as directed.'

10. **Which of the following auxiliary labels are required on all creams prepared extemporaneously?**
a. Discard date of 4 weeks and 'Not to be taken'.
b. Discard date of 4 weeks and 'For external use only'.
c. Discard date of 2 weeks and 'For external use only'.
d. Discard date of 2 weeks and 'Not to be taken'.

11. **Why is hygiene particularly important when preparing creams?**

12. **What is a suitable container for a cream?**

Formulation questions

This section contains details of extemporaneous products to be made in the same way as the examples earlier in this chapter. For each example, provide answers using the following sections:

1. **Use of the product**
2. **Is it safe and suitable for the intended purpose?**
3. **Calculation of formula for preparation**
4. **Method of preparation**
   a. Solubility where applicable
   b. Vehicle/diluent
   c. Preservative
   d. Flavouring when appropriate
5. **Choice of container**
6. **Labelling considerations**
   a. Title
   b. Quantitative particulars
   c. Product-specific cautions (or additional labelling requirements)
   d. Directions to patient – interpretation of Latin abbreviations where necessary
   e. Recommended *British National Formulary* cautions when suitable
   f. Discard date
   g. Sample label (you can assume that the name and address of the pharmacy and the words 'Keep out of the reach and sight of children' are pre-printed on the label)
7. **Advice to patient.**

13. You receive a prescription in your pharmacy with the following details:

Patient:	Mrs Lily Evans, 4 Park Street, Astonbury
Age:	78
Prescription:	Dimeticone Cream BPC
Directions:	Use as a barrier cream after bathing
Mitte:	30 g

14. You receive a prescription in your pharmacy with the following details:

Patient:	Mrs Avril Asker, 12 Bishop Road, Astonbury	
Age:	35	
Prescription:	Zinc oxide	15%
	Calamine	5%
	In aqueous cream	
Directions:	Apply tds mdu	
Mitte:	20 g	

chapter 6
Ointments, pastes and gels

Overview

Upon completion of this chapter, you should be able to:
- understand the difference between ointments, pastes and gels
- prepare an ointment from first principles
- incorporate solids and liquids into an ointment base
- select an appropriate container in which to package ointments, pastes and gels
- prepare an appropriate label for ointments, pastes and gels.

Introduction and overview of ointments, pastes and gels

This chapter will focus on the following dosage forms:
- ointments
- pastes
- gels.

Ointments
British Pharmacopoeia (BP) definition
Ointments are formulated to provide preparations that are immiscible, miscible or emulsifiable with the skin secretion. Hydrophobic ointments and water-emulsifying ointments are intended to be applied to the skin or certain mucous membranes for emollient, protective, therapeutic or prophylactic purposes where a degree of occlusion is desired. Hydrophilic ointments are miscible with the skin secretion and are less emollient as a consequence.

Pastes
Pastes are semi-solid preparations for external use. They consist of finely powdered medicaments combined with White Soft Paraffin BP or Liquid Paraffin BP or with a non-greasy base made from glycerol, mucilages or soaps. Pastes contain a high proportion of powdered ingredients and therefore are normally very stiff. Because pastes are stiff they do not spread easily and therefore this localises drug delivery. This is particularly important if the ingredient to be applied to the skin is corrosive such as dithranol, coal tar or salicylic acid. It is easier to apply a paste to a discrete

Definition

Ointments are preparations for external application but differ from creams in that they have greasy bases. The base is usually anhydrous and therefore most ointments are immiscible with skin secretions. Ointments usually contain a medicament or a mixture of medicaments dissolved or dispersed in the base.

skin area such as a particular lesion or plaque and not therefore compromise the integrity of healthy skin.

Pastes are also useful for absorbing harmful chemicals such as the ammonia which is released by bacterial action on urine and so are often used in nappy products. Also, because of their high powder content, they are often used to absorb wound exudates.

Because pastes are so thick they can form an unbroken layer over the skin which is opaque and can act as a sun filter. This makes them suitable for use for skiers as they prevent excessive dehydration of the skin (wind burn) in addition to sun blocking.

The principal use of pastes was traditionally as an antiseptic, protective or soothing dressing. Often before application the paste was applied to lint and then applied as a dressing.

Gels

Pharmaceutical gels are often simple phase, transparent semi-solid systems that are being increasingly used as pharmaceutical topical formulations. The liquid phase of the gel may be retained within a three-dimensional polymer matrix. Drugs can be suspended in the matrix or dissolved in the liquid phase.

Advantages of gels

- Stable over long periods of time
- Good appearance
- Suitable vehicles for applying medicaments to skin and mucous membranes giving high rates of release of the medicament and rapid absorption.

Gels are usually translucent or transparent and have a number of uses:

- Anaesthetic gels
- Coal tar gels for use in treatment of psoriasis or eczema
- Lubricant gels
- Spermicidal gels.

General method

This section contains information on the preparation of ointments by fusion and the incorporation of both solids and liquids into ointment bases.

Fusion

This involves melting together the bases over a water bath (see Figure 5.1) before incorporating any other ingredients. The ointment base may include a mixture of waxes, fats and oils, of which some are solid at room temperature and others are liquid.

- Hard: Paraffin BP, Beeswax BP, Cetostearyl Alcohol BP
- Soft: Yellow and White Soft Paraffin BP, Wool Fat BP
- Liquid: Liquid Paraffin BP and vegetable oils.

General method (fusion)
1. Always make excess as transference losses will always occur.
2. Determine the melting points of the fatty bases and then melt together. Starting with the base with the highest melting point, each base should be melted at the lowest possible temperature as the mixture progressively cools.
3. Add the ingredients to an evaporating basin over a water bath (see Figure 5.1) to avoid overheating – use a thermometer to check the temperature regularly.
4. As the first base cools, add the ingredients with decreasing melting points at the respective temperatures, stirring continuously to ensure a homogeneous mix before leaving to set. It is important to stir gently to avoid incorporating excess air, which could result in localised cooling and a lumpy product.

See Ointments video for a demonstration of the fusion method.

The incorporation of powders into an ointment base
- **Soluble solids** should be added to the molten fatty bases at the lowest possible temperature and the mixture stirred until cold. Alternatively, if using a pre-prepared base, soluble solids may be incorporated using the method employed for insoluble solids.
- **Insoluble solids** should be incorporated using a glass tile and spatula (see Figure 5.2). If there is more than one powder to be added, these should be mixed in a mortar using the 'doubling-up' method.
- **Coarse powders.** A minimum quantity of molten fatty base should be placed in the centre of the glass tile and used to levigate the powders. A considerable shearing force should be applied to avoid a gritty product.

  The powder/fatty base mixture may then either be returned to the evaporating basin with the remaining fatty base and stirred until cold or the remaining fatty base in the evaporating basin may be allowed to cool and triturated with the powder/fatty base mixture on the tile.
- **Fine powders** may be triturated into the otherwise finished ointment on a glass tile. Small amounts of powder should be added to an equal amount of ointment, i.e. using the 'doubling-up' technique. These should be well triturated to incorporate all of the ointment base. Alternatively, a small amount of powder may be levigated with some molten

ointment base on a tile and the resulting mixture returned to the remaining molten mass and stirred to achieve a homogeneous product

See Ointments video for a demonstration of the incorporation of a powder into a base.

The incorporation of liquids into an ointment base

- **Non-volatile, miscible liquids** may be mixed with the molten fat in the evaporating basin. Alternatively, if a pre-prepared base is used, then incorporate as for volatile or immiscible liquids.
- **Volatile or immiscible liquids**, e.g. coal tar solutions, should be triturated with the ointment on the glass tile.

A very small amount of the ointment should be placed on the glass tile and a 'well' made in the centre. Traditionally, small quantities of liquid should be gently folded in to avoid splashing. An alternative method is to spread a small amount of the ointment on the tile and then score it with a spatula. Then add small quantities of the liquid and fold into the base gently.

If using coal tar or other volatile ingredients, these should not be weighed until immediately before use and the beaker in which it has been weighed should be covered with a watch glass to prevent evaporation. In addition, always remember that volatile ingredients should not be added to molten bases.

See Ointments video for a demonstration of the incorporation of a liquid into a base.

Worked examples

Example 6.1
The preparation of Simple Ointment BP
You receive a prescription in your pharmacy with the following details:

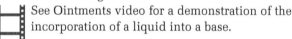

Patient:	Mr Martin Ally, 5 Longmeadow, Astonbury
Age:	56
Prescription:	Simple Ointment BP
Directions:	Mdu
Mitte:	30 g

1. **Use of the product**
    The product is used as an emollient (*Martindale* 33rd edn, p 1408).
2. **Is it safe and suitable for the intended purpose?**
    This is an official formula for an emollient, therefore the formula and frequency of application are safe.

3. **Calculation of formula for preparation**
    Prepare 30 g Simple Ointment BP.

Product formula
(from the *British Pharmacopoeia* 1988, p 713)

	Master	100 g	10 g	40 g
Cetostearyl Alcohol BP	50 g	5 g	0.5 g	2 g
Hard Paraffin BP	50 g	5 g	0.5 g	2 g
Wool Fat BP	50 g	5 g	0.5 g	2 g
White/Yellow Soft Paraffin BP	850 g	85 g	8.5 g	34 g

4. **Method of preparation**
a. Solubility where applicable
    Not applicable.
b. Vehicle/diluents
    White/Yellow Soft Paraffin BP is used as
    the diluent as per the product formula.
c. Preservative
    There is no preservative included as per
    the product formula.
d. Flavouring when appropriate
    Ointments are for external use and so no
    flavouring is required

The following method would be used to
prepare 40 g of Simple Ointment BP from
the formula above:
Note that the melting points of the
ingredients are:
Cetostearyl Alcohol BP: 49–56°C (*British
Pharmacopoeia* 1988, p 111)
Hard Paraffin BP: 50–61°C (*British
Pharmacopoeia* 1988, p 415)
White/Yellow Soft Paraffin BP: 38–56°C (*British
Pharmacopoeia* 1988, p 416)
Wool Fat BP: 38–44°C (*British Pharmacopoeia* 1988, p 601)

1. Weigh 2 g Hard Paraffin BP on a Class II or electronic
    balance.
2. Weigh 2 g Cetostearyl Alcohol BP on a Class II or electronic
    balance.
3. Weigh 2 g Wool Fat BP on a Class II or electronic balance.
4. Weigh 34 g Yellow/White Soft Paraffin BP on a Class II or
    electronic balance.
5. Place the Hard Paraffin BP into an evaporating dish and
    melt over a water bath.
6. Remove from the heat and add the other ingredients in
    descending order of melting point until all are melted in

Tips

The exact quantity cannot be
prepared for an ointment: a
suitable overage must be produced
in order to dispense the required
amount.
    Yellow or White Soft Paraffin
BP may be used when making this
ointment, which is often used as
a base for other ointments. As
a general rule, if it is to be used
as a base and the ingredients
to be added are coloured (e.g.
Coal Tar Solution BP), Yellow Soft
Paraffin BP would be used. If
the ingredients to be added are
white or pale in colour (e.g. Zinc
Oxide BP or Calamine BP), White
Soft Paraffin BP would be used to
produce a more pharmaceutically
elegant product.

(return to the heat if necessary to ensure even melting but take care not to overheat).

7. Stir until cold.
8. Weigh 30 g and pack into a collapsible tube or amber glass jar. Label and dispense.

5. **Choice of container**

   A collapsible tube or plain amber jar would be most suitable.

6. **Labelling considerations**

   a. Title

   The product is official, therefore the following title would be suitable: 'Simple Ointment BP'

   b. Quantitative particulars

   Quantitative particulars are not required as the product is official.

   c. Product-specific cautions (or additional labelling requirements)

   'For external use only' will need to be added to the label as the product is an ointment for external use.

   d. Directions to patient – interpretation of Latin abbreviations where necessary

   'Apply as directed.'

   e. Recommended *British National Formulary* cautions when suitable

   Not applicable.

   f. Discard date

   The product is an ointment and so will attract a 3-month discard date.

   g. Sample label (you can assume that the name and address of the pharmacy and the words 'Keep out of the reach and sight of children' are pre-printed on the label):

Simple Ointment BP	30 g
Apply as directed.	
For external use only	
Do not use after (3 months)	
Mr Martin Ally	Date of dispensing

7. **Advice to patient**

   The patient would be advised to apply the ointment as directed. In addition, the discard date and the fact that the product is for external use only would be highlighted to the patient. If further direction for use is requested, the patient could be advised to apply the ointment frequently as the product is an emollient.

Example 6.2
The preparation of Calamine and Coal Tar Ointment BP

You receive a prescription in your pharmacy with the following details:

Patient:	Mr Kenny Doyle, 13 Acres Street, Astonbury
Age:	38
Prescription:	Calamine and Coal Tar Ointment BP
Directions:	Apply once or twice daily
Mitte:	20 g

1. **Use of the product**
   The product is used to treat psoriasis and occasionally chronic atopic eczema (*British National Formulary* 61st edn, p 718).
2. **Is it safe and suitable for the intended purpose?**
   This is an official preparation, therefore the formula is safe and suitable for purpose. The *British National Formulary* (61st edn, p 718) suggests an application frequency of 1–2 times daily, which corresponds to the prescribed frequency.
3. **Calculation of formula for preparation**
   Prepare 20 g Calamine and Coal Tar Ointment BP.

Product formula
(from the *British Pharmacopoeia* 2007, p 2376):

	Master	100 g	10 g	30 g
Calamine BP	125 g	12.5 g	1.25 g	3.75 g
Strong Coal Tar Solution BP	25 g	2.5 g	0.25 g	0.75 g
Zinc Oxide BP	125 g	12.5 g	1.25 g	3.75 g
Hydrous Wool Fat BP	250 g	25 g	2.5 g	7.5 g
White Soft Paraffin BP	475 g	47.5 g	4.75 g	14.25 g

4. **Method of preparation**
a. Solubility where applicable
   Not applicable.
b. Vehicle/diluent
   White Soft Paraffin BP is used as the diluent as per the product formula.
c. Preservative
   There is no preservative included as per the product formula.
d. Flavouring when appropriate
   Ointments are for external use and so no flavouring is required

Tips

The exact quantity cannot be prepared for an ointment, a suitable overage must be produced in order to dispense the required amount.

Note that the quantity of Strong Coal Tar Solution BP is in grams and so must be weighed.

The following method would be used to prepare 30 g of Calamine and Coal Tar Ointment BP from the formula above. Note that the melting points of the ingredients are as follows: Hydrous Wool Fat BP: 38–44°C (*British Pharmacopoeia* 1988, p 602).

White/Yellow Soft Paraffin BP: 38–56°C (*British Pharmacopoeia* 1988, p 416).

1. Weigh 3.75 g Calamine BP on a Class II or electronic balance.
2. Weigh 3.75 g Zinc Oxide BP on a Class II or electronic balance.
3. Transfer the Calamine BP and the Zinc Oxide BP to a porcelain mortar and triturate together with a pestle.
4. Weigh 14.25 g White Soft Paraffin BP on a Class II or electronic balance.
5. Weigh 7.5 g Hydrous Wool Fat BP on a Class II or electronic balance.
6. Place the White Soft Paraffin BP into an evaporating dish and melt over a water bath.
7. Remove from the heat and add the Hydrous Wool Fat BP. Stir until melted to ensure an even well-mixed base.
8. Transfer the powders to a glass tile and levigate with some of the molten base.
9. Transfer the powder/base mix to the rest of the molten base and stir until homogeneous.
10. Weigh 0.75 g Strong Coal Tar Solution BP on a Class II or electronic balance.
11. Allow the base/powder mixture to cool and add the Strong Coal Tar Solution BP and stir until homogeneous.
12. Weigh 20 g of the product and pack into a collapsible tube or amber glass jar. Label and dispense.

Tips

The Strong Coal Tar Solution BP cannot be added until the bases are quite cool (less than 40°C) as it is a volatile preparation. This method would also avoid heating the Strong Coal Tar Solution BP and therefore reduce the volatilisation of some of the coal tar constituents and reduce the risk of sedimentation.

5. **Choice of container**
   A collapsible tube or plain amber jar would be most suitable.
6. **Labelling considerations**
a. Title
   The product is official therefore the following title would be suitable: 'Calamine and Coal Tar Ointment BP'.
b. Quantitative particulars
   Quantitative particulars are not required as the product is official.
c. Product-specific cautions (or additional labelling requirements)
   'For external use only' will need to be added to the label as the product is an ointment for external use. In addition, the

product contains coal tar and so the following warning should be added to the label: 'Caution: may stain hair, skin and fabrics'.

d. Directions to patient – interpretation of Latin abbreviations where necessary
'Apply ONCE or TWICE a day.'

e. Recommended *British National Formulary* cautions when suitable
Not applicable.

f. Discard date
The product is an ointment and so will attract a 3-month discard date.

g. Sample label (you can assume that the name and address of the pharmacy and the words 'Keep out of the reach and sight of children' are pre-printed on the label):

Calamine and Coal Tar Ointment BP	**20 g**
Apply ONCE or TWICE a day.	
For external use only	
Caution: may stain hair, skin and fabrics	
Do not use after (3 months)	
Mr Kenny Doyle	Date of dispensing

7. **Advice to patient**
The patient would be advised to apply the ointment once or twice a day. In addition, the discard date, the fact that the product is for external use only and that it may stain hair, skin and fabrics would be highlighted to the patient.

Example 6.3
The preparation of Zinc Ointment BP

You receive a prescription in your pharmacy with the following details:

Patient:	Master John Pike, 234 High Street, Astonbury
Age:	2
Prescription:	Zinc Ointment BP
Directions:	paa qds
Mitte:	20 g

1. **Use of the product**
For nappy and urinary rash and eczematous conditions (*British National Formulary* 61st edn, p 705).

2. **Is it safe and suitable for the intended purpose?**
This is an official preparation, therefore the formula is safe and suitable for purpose. External products which do not

contain potent ingredients are normally applied liberally when required. Therefore, an application frequency of four times a day would be suitable for the treatment of nappy rash.

3. **Calculation of formula for preparation**
    Prepare 20 g of Zinc Ointment BP.

Product formula
(from the *British Pharmacopoeia* 2007, p 2992)

	Master	100 g	10 g	30 g
Zinc Oxide BP	150 g	15 g	1.5 g	4.5 g
Simple Ointment BP	850 g	85 g	8.5 g	25.5 g

Tips

The exact quantity cannot be prepared for an ointment as losses will be experienced on transfer. Therefore, a suitable overage must be produced in order to dispense the required amount.

Tips

The Zinc Oxide BP is transferred to a mortar so that the size of any lumps can be reduced enabling a smooth product to be made.

Tips

To triturate means to mix. A smooth product will be produced if the Zinc Oxide BP is finely sifted and then just mixed with the base. The particle size reduction of the Zinc Oxide BP in this example has only been achieved by mixing in a mortar, rather than sifting with a sieve. In order to achieve a smooth product, considerably more lateral shearing force will need to be applied to the powder and we suggest that the process employed would be more akin to levigation (wet grinding), but in this case using a semi-solid base rather than a molten base.

4. **Method of preparation**
a. Solubility where applicable
    Not applicable.
b. Vehicle/diluent
    Simple Ointment BP is used as the diluent as per the product formula.
c. Preservative
    There is no preservative included as per the product formula.
d. Flavouring when appropriate
    Ointments are for external use and so no flavouring is required.

The following method would be used to prepare 30 g of Zinc Ointment BP from the formula above:

1. Weigh 4.5 g Zinc Oxide BP on a Class II or electronic balance.
2. Transfer to a porcelain mortar and stir with a pestle.
3. Transfer the Zinc Oxide BP to a glass tile.
4. Weigh 25.5 g Simple Ointment BP on a Class II or electronic balance.
5. Transfer the Simple Ointment BP to the glass tile.
6. Triturate the Zinc Oxide BP with the Simple Ointment BP until a smooth product is formed.
7. Weigh 20 g of the product and pack into a collapsible tube or amber glass jar. Label and dispense.

5. **Choice of container**
    A collapsible tube or plain amber jar would be most suitable.
6. **Labelling considerations**
a. Title
    The product is official, therefore the following title would be
    suitable: 'Zinc Ointment BP'.
b. Quantitative particulars
    Quantitative particulars are not required as the product is official.
c. Product-specific cautions (or additional labelling requirements)
    'For external use only' will need to be added to the label as the
    product is an ointment for external use.
d. Directions to patient – interpretation of Latin abbreviations
    where necessary
    'Apply to the affected areas FOUR times a day.'
e. Recommended *British National Formulary* cautions when
    suitable
    Not applicable.
f. Discard date
    The product is an ointment and so will attract a 3-month
    discard date.
g. Sample label (you can assume that the name and address of the
    pharmacy and the words 'Keep out of the reach and sight of
    children' are pre-printed on the label):

Zinc Ointment BP	**20 g**
Apply to the affected areas FOUR times a day.	
For external use only	
Do not use after (3 months)	
Master John Pike	Date of dispensing

7. **Advice to patient**
    The parent/guardian would be advised to apply the ointment
    to the affected areas four times a day. In addition, the discard
    date and the fact that the product is for external use only
    would be highlighted to the parent/guardian.

Example 6.4
Unofficial ointment request from local doctor

You receive a prescription in your pharmacy with the following
details:

Patient:	Mr David Raymond, 13 Gas Street, Astonbury
Age:	38
Prescription:	CCS & S Ointment
Directions:	Apply mdu
Mitte:	20 g

1. **Use of the product**
   The product is used to treat psoriasis (*Martindale* 35th edn, pp 1451 and 1452).

2. **Is it safe and suitable for the intended purpose?**
   This is an unofficial formula, therefore the formula and frequency of application need to be checked. The formula has originated from a specialist dermatological clinic (see below) and they advise a frequency of application for CCS & S ointment of once to twice daily depending on severity of the condition being treated. The product is therefore safe and suitable for use.

3. **Calculation of formula for preparation**
   Prepare 20 g of CCS & S Ointment.
   The local dermatology clinic has a formulary which gives the following formula:

Salicylic Acid BP	3%
Camphor BP	3%
Sulphur BP	3%
Phenol BP	3%
White Soft Paraffin BP	to 100%

Product formula

	100 g	10 g	30 g
Salicylic Acid BP	3 g	300 mg	900 mg
Camphor BP	3 g	300 mg	900 mg
Sulphur BP	3 g	300 mg	900 mg
Phenol BP	3 g	300 mg	900 mg
White Soft Paraffin BP	88 g	8.8 g	26.4 g

4. **Method of preparation**
   a. Solubility where applicable
      A liquid is formed when Camphor BP and Phenol BP are mixed together (*Martindale* 35th edn, p 2064).
   b. Vehicle/diluent
      White Soft Paraffin BP is used as the diluent as per the product formula.
   c. Preservative
      Phenol BP would act as a preservative in addition to acting as an antiseptic.
   d. Flavouring when appropriate
      Ointments are for external use and so no flavouring is required.

   The following method would be used to prepare 30 g of CCS & S Ointment from the formula above.

Note that when phenol is combined with camphor, a liquid mixture results.

1. Weigh 900 mg Salicylic Acid BP on a Class II or electronic balance.
2. Weigh 900 mg Sulphur BP on a Class II or electronic balance.
3. Transfer the Salicylic Acid BP to a glass mortar and grind with a pestle to reduce particle size.
4. Add the Sulphur BP and continue mixing.
5. Transfer the mixed powders to a glass tile.
6. Weigh 26.4 g White Soft Paraffin BP on a Class II or electronic balance.
7. Transfer the White Soft Paraffin BP to the glass tile.
8. Triturate the powders with the White Soft Paraffin BP until a smooth product.
9. Weigh 900 mg of Camphor BP on a Class II or electronic balance.
10. Transfer to a clean dry glass mortar.
11. Weigh 900 mg of Phenol BP on a Class II or electronic balance.
12. Add the Phenol BP to the Camphor BP and mix together with a pestle.
13. Make a well in the ointment mass and add the liquid mixture.
14. Triturate until all the liquid is incorporated and a homogeneous product is formed.
15. Weigh 20 g of product and pack into a collapsible tube or amber glass jar and label.

Tips

Note that a vulcanite spatula would be the spatula of choice as traditional stainless steel spatulas may react with acids, tannins, iodine and mercury salts, etc.
  Vulcanite (also called ebonite) is a hard, usually black, rubber produced by vulcanising natural rubber with sulphur. Such spatulas are used for making ointments containing corrosive substances or substances that react with steel.

Tips

The Camphor BP and Phenol BP are weighed and mixed at the final stage of preparation of the product as both are volatile ingredients.

5. **Choice of container**

A collapsible tube or plain amber jar would be most suitable.

6. **Labelling considerations**

a. Title

The product is unofficial, therefore the following title would be suitable: 'CCS & S Ointment'.

b. Quantitative particulars

Quantitative particulars are required as the product is unofficial. As the product is for external use, the quantitative particulars would be expressed per container:

The product contains:

Salicylic Acid BP	3%
Camphor BP	3%
Sulphur BP	3%
Phenol BP	3%
White Soft Paraffin BP	to 100%

c.   Product-specific cautions (or additional labelling requirements)
     'For external use only' will need to be added to the label as the
     product is an ointment for external use.
d.   Directions to patient – interpretation of Latin abbreviations
     where necessary
     'Apply as directed.'
e.   Recommended *British National Formulary* cautions when
     suitable
     Not applicable.
f.   Discard date
     The product is an ointment and so will attract a 3-month
     discard date.
g.   Sample label (you can assume that the name and address of the
     pharmacy and the words 'Keep out of the reach and sight of
     children' are pre-printed on the label):

CCS & S Ointment		**20 g**
	Apply as directed.	
	For external use only	
	Do not use after (3 months)	
The product contains:		
Salicylic Acid BP	3%	
Camphor BP	3%	
Sulphur BP	3%	
Phenol BP	3%	
White Soft Paraffin BP	to 100%	
Mr David Raymond		Date of dispensing

7.   **Advice to patient**
     The patient would be advised to apply the ointment as
     directed. In addition, the discard date and the fact that the
     product is for external use only would be highlighted to the
     patient. If further direction for use is requested, the patient
     could be advised to apply the cream once or twice a day as,
     although the concentration of Salicylic Acid BP is quite low
     when compared to preparations used in the treatment of warts
     and hard skin, it will act as a keratolytic and so a maximum
     application frequency of once or twice a day would be
     appropriate.

Example 6.5
Unofficial ointment request from local doctor

You receive a prescription in your pharmacy with the following details:

Patient:	Ms Daphne Stokes, 74 Fish Street, Astonbury
Age:	42
Prescription:	Salicylic Acid 2% in Betnovate Ointment
Directions:	Apply mdu
Mitte:	40 g

1. **Use of the product**
   Used to treat psoriasis (Salicylic Acid: *Martindale* 35th edn, p 1451; Betnovate Ointment: *British National Formulary* 61st edn, p 710).
2. **Is it safe and suitable for the intended purpose?**
   The suitability of the product formula must be decided by the compounder. Refer to a diluent directory which in this case states that up to 5% of Salicylic Acid BP can be added to Betnovate Ointment. Had the formula requested Betnovate Cream, this would have been unsuitable as the Salicylic Acid BP causes the cream to crack. The suggested frequency of application for Betnovate Ointment is to apply thinly 1–2 times daily (*British National Formulary* 61st edn, p 710). Therefore, the product is safe and suitable for use.
3. **Calculation of formula for preparation**
   Prepare 40 g of Salicylic Acid BP 2% in Betnovate Ointment. Note: Suitable sources to provide information on the dilution of creams and ointments would be:
   - National Pharmacy Association Diluent Directory
   - product data sheet (Summary of Product Characteristics – SPC)
   - reports in the pharmaceutical literature
   - personal contact with product manufacturer.

Product formula

	Master	100 g	10 g	50 g
Salicylic Acid BP	2%	2 g	200 mg	1 g
Betnovate Ointment	98%	98 g	9.8 g	49 g

4. **Method of preparation**
   a. Solubility where applicable
      Not applicable.
   b. Vehicle/diluent
      Betnovate Ointment is used as the base as per the product formula.

c. **Preservative**
    The product contains no additional preservative (apart
    from any preservative already present within the Betnovate
    Ointment).
d. **Flavouring when appropriate**
    Ointments are for external use and so no flavouring is required.

The following method would be used to prepare 50 g of
Salicylic Acid BP 2% in Betnovate Ointment from the formula
above:

1. Weigh 1 g Salicylic Acid BP on a Class II or electronic
    balance.
2. Transfer to a glass mortar and grind with a pestle to reduce
    any lumps in the powder.
3. Transfer the powder to a glass tile.
4. Weigh 49 g Betnovate Ointment.
5. Transfer to the tile.
6. Triturate the Salicylic Acid BP and the Betnovate Ointment
    together, remembering the 'doubling-up' technique for
    adequate mixing.
7. Weigh 40 g of the resultant ointment, pack into a
    collapsible tube or amber glass jar, label and dispense.

5. **Choice of container**
    A collapsible tube or plain amber jar would be most suitable.
6. **Labelling considerations**
a. Title
    The product is unofficial, therefore the following title would
    be suitable: 'Salicylic Acid 2% in Betnovate Ointment'.
b. Quantitative particulars
    Quantitative particulars are required as the product is
    unofficial. As the product is for external use, the quantitative
    particulars would be expressed per container:

    The product contains:

Salicylic Acid BP	2%
Betnovate Ointment	98%

c. Product-specific cautions (or additional labelling requirements)
    'For external use only' will need to be added to the label as the
    product is an ointment for external use.
d. Directions to patient – interpretation of Latin abbreviations
    where necessary
    'Apply as directed.'
e. Recommended *British National Formulary* cautions when
    suitable

The *British National Formulary* (61st edn, p 710) recommends the following caution:

Label 28 – 'Spread thinly on the affected skin only'.

f. Discard date

The product is an ointment and so would normally attract a 3-month discard date. However, as the product is a diluted proprietary ointment, it is common to assign a shorter 2-week discard date.

g. Sample label (you can assume that the name and address of the pharmacy and the words 'Keep out of the reach and sight of children' are pre-printed on the label):

Salicylic Acid 2% in Betnovate Ointment	**40 g**

Apply as directed.

Spread thinly on the affected skin only

For external use only

Do not use after (2 weeks)

The product contains:

Salicylic Acid BP	2%
Betnovate Ointment	98%
Ms Daphne Stokes	Date of dispensing

7. **Advice to patient**

The patient would be advised to apply the ointment as directed. In addition, the discard date and the fact that the product is for external use only would be highlighted to the patient. If further direction for use is requested, as the preparation contains Betnovate Ointment the patient could be advised to apply the ointment thinly 1–2 times daily (*British National Formulary* 61st edn, p 710). In addition, although the concentration of Salicylic Acid BP is quite low when compared to preparations used in the treatment of warts and hard skin, it will act as a keratolytic and so a maximum application frequency of once or twice a day would also be appropriate.

Example 6.6
The preparation of Compound Zinc Paste BP

You receive a prescription in your pharmacy with the following details:

Patient:	Mrs Sandra Jones, 6 Summet Drive, Astonbury
Age:	56
Prescription:	Compound Zinc Paste BP
Directions:	paa tds
Mitte:	20 g

1. **Use of the product**
   Mild astringent (*Martindale* 35th edn, p 1458).
2. **Is it safe and suitable for the intended purpose?**
   This is an official preparation, therefore the formula is safe and suitable for purpose.
3. **Calculation of formula for preparation**
   Prepare 20 g Compound Zinc Paste BP.

Product formula
(from the *British Pharmacopoeia* 2007, p 2993)

	Master	100 g	10 g	30 g
Zinc Oxide BP	250 g	25 g	2.5 g	7.5 g
Starch BP	250 g	25 g	2.5 g	7.5 g
White Soft Paraffin BP	500 g	50 g	5 g	15 g

4. **Method of preparation**
a. Solubility where applicable
   Not applicable.
b. Vehicle/diluent
   White Soft Paraffin BP is used as the diluent as per the product formula.
c. Preservative
   There is no preservative included as per the product formula.
d. Flavouring when appropriate
   Pastes are for external use and so no flavouring is required.

The following method would be used to prepare 30 g of Compound Zinc Paste BP from the formula above. Note the melting point of the base:
White/Yellow Soft Paraffin BP 38–56°C (*British Pharmacopoeia* 1988, p 416).

1. Weigh 7.5 g Zinc Oxide BP on a Class II or electronic balance.
2. Weigh 7.5 g Starch BP on a Class II or electronic balance.
3. Weigh 15 g White Soft Paraffin BP on a Class II or electronic balance.
4. Transfer the Zinc Oxide BP to a porcelain mortar.
5. Add the Starch BP to the mortar and triturate with the pestle to form an evenly mixed powder.
6. Transfer the powder to a glass tile.
7. Transfer the White Soft Paraffin BP to the glass tile.
8. Mix the powders with the White Soft Paraffin BP using a metal spatula and remembering the principle of 'doubling-up'.

Tips

A porcelain mortar is used because of the volume of powder involved.

9. Triturate until a smooth product is formed.

10. Weigh 20 g and pack into a collapsible tube or amber glass jar. Label and dispense.

Tips

An alternative way to prepare this paste would involve melting the base then combining the powders with the molten base and stirring until cooled.

5. **Choice of container**

A collapsible tube or plain amber jar would be most suitable.

6. **Labelling considerations**

a. Title

The product is official, therefore the following title would be suitable: 'Compound Zinc Paste BP'.

b. Quantitative particulars

Quantitative particulars are not required as the product is official.

c. Product-specific cautions (or additional labelling requirements)

'For external use only' will need to be added to the label as the product is a paste for external use.

d. Directions to patient – interpretation of Latin abbreviations where necessary

'Apply to the affected areas THREE times a day.'

e. Recommended *British National Formulary* cautions when suitable

Not applicable.

f. Discard date

The product is a paste and so will attract a 3-month discard date.

g. Sample label (You can assume that the name and address of the pharmacy and the words 'Keep out of the reach and sight of children' are pre-printed on the label):

Compound Zinc Paste BP	20 g
Apply to the affected areas THREE times a day.	
For external use only	
Do not use after (3 months)	
Mrs Sandra Jones	Date of dispensing

7. **Advice to patient**

The patient would be advised to apply the ointment to the affected areas three times a day. In addition, the discard date and the fact that the product is for external use only would be highlighted to the patient.

Example 6.7
The preparation of Zinc and Coal Tar Paste BP

You receive a prescription in your pharmacy with the following details:

Patient:	Mr Scott Bird, 24 Fleet Drive, Astonbury
Age:	38
Prescription:	Zinc and Coal Tar Paste BP
Directions:	Apply bd mdu
Mitte:	20 g

1. **Use of the product**
   Used to treat psoriasis (and occasionally chronic atopic eczema) (*British National Formulary* 61st edn, p 718).
2. **Is it safe and suitable for the intended purpose?**
   This is an official preparation, therefore the formula is safe and suitable for purpose. The *British National Formulary* (61st edn, p 718) suggests an application frequency of 1–2 times daily which is consistent with the prescription.
3. **Calculation of formula for preparation**
   Prepare 20 g Zinc and Coal Tar Paste BP.

Product formula
(from the *British Pharmacopoeia* 2007, page 2993)

	Master	100 g	10 g	30 g
Emulsifying Wax BP	50 g	5 g	500 mg	1.5 g
Coal Tar BP	60 g	6 g	600 mg	1.8 g
Zinc Oxide BP	60 g	6 g	600 mg	1.8 g
Starch BP	380 g	38 g	3.8 g	11.4 g
Yellow Soft Paraffin BP	450 g	45 g	4.5 g	13.5 g

4. **Method of preparation**
a. Solubility where applicable
   Not applicable.
b. Vehicle/diluent
   Yellow Soft Paraffin BP is used as the diluent as per the product formula.
c. Preservative
   There is no preservative included as per the product formula.
d. Flavouring when appropriate
   Pastes are for external use and so no flavouring is required.

The following method would be used to prepare 30 g of Zinc and Coal Tar Paste BP from the formula above.

Note the melting points of the ingredients:

Emulsifying Wax BP   52°C

White/Yellow Soft Paraffin BP   38–56°C

1. Weigh 1.8 g Zinc Oxide BP on a Class II or electronic balance.
2. Transfer to a porcelain mortar.
3. Weigh 11.4 g Starch BP on a Class II or electronic balance.
4. Add the Starch BP to the Zinc Oxide BP in the porcelain mortar and stir with the pestle.
5. Weigh 1.5 g Emulsifying Wax BP on a Class II or electronic balance.
6. Weigh 1.8 g Coal Tar BP on a Class II or electronic balance.
7. Weigh 13.5 g Yellow Soft Paraffin BP on a Class II or electronic balance.
8. Place the Emulsifying Wax BP into an evaporating dish and melt over a water bath at 70°C.
9. Add the Coal Tar BP and half of the Yellow Soft Paraffin BP to the evaporating basin.
10. Stir at 70°C until melted.
11. Add the remaining Yellow Soft Paraffin BP stir until melted.
12. Cool to approximately 30°C and add the powders and stir constantly until cold.
13. Weigh 20 g of the paste transfer to a collapsible tube or amber glass jar, dispense and label.

5. **Choice of container**

    A collapsible tube or plain amber jar would be most suitable.

6. **Labelling considerations**

a. Title

    The product is official, therefore the following title would be suitable: 'Zinc and Coal Tar Paste BP'.

b. Quantitative particulars

    Quantitative particulars are not required as the product is official.

c. Product-specific cautions (or additional labelling requirements)

    'For external use only' will need to be added to the label as the product is a paste for external use.

Tips

The powders must be mixed, remembering the principle of 'doubling-up' in order to ensure even mixing of the powders.

Tips

The above method is as recommended by the *British Pharmacopoeia*. An alternative method would be:

- Melt the Yellow Soft Paraffin BP and Emulsifying Wax BP together at the lowest possible temperature, stirring until cool, to make a homogeneous product.
- Mix the powders as before but transfer them to a glass tile and incorporate the powders into the base using a spatula.
- Finally, using a spatula (preferably ebonite), incorporate the Coal Tar BP.
- This method may be preferred because of the possible problem of toxicity associated with Coal Tar BP. This method would avoid heating the Coal Tar BP and therefore reduce the volatilisation of some of the coal tar constituents and reduce the risk of sedimentation.

d. Directions to patient – interpretation of Latin abbreviations where necessary
    'Apply TWICE a day as directed.'

e. Recommended *British National Formulary* cautions when suitable
    Not applicable.

f. Discard date
    The product is a paste and so will attract a 3-month discard date.

g. Sample label (you can assume that the name and address of the pharmacy and the words 'Keep out of the reach and sight of children' are pre-printed on the label):

Zinc and Coal Tar Paste BP	**20 g**
Apply TWICE a day as directed.	
For external use only	
Do not use after (3 months)	
Mr Scott Bird	Date of dispensing

7. **Advice to patient**
    The patient would be advised to apply the ointment twice a day as directed. In addition, the discard date and the fact that the product is for external use only would be highlighted to the patient.

Example 6.8
The preparation of Dithranol Paste BP

You receive a prescription in your pharmacy with the following details:

Patient:	Miss Amy Smith, 12 Ash Drive, Astonbury
Age:	30
Prescription:	Dithranol Paste BP 0.1%
Directions:	Apply mdu
Mitte:	40 g

1. **Use of the product**
    Used in treatment of subacute and chronic psoriasis (*British National Formulary* 61st edn, p 719).

2. **Is it safe and suitable for the intended purpose?**
    This is an official preparation, therefore the formula is safe and suitable for purpose.

3. **Calculation of formula for preparation**
    Prepare 40 g of Dithranol Paste BP 0.1%.
    (Note: the strengths of dithranol paste can vary between 0.1% and 1%.)

Product formula
(from the *British Pharmacopoeia* 2007, p 2525)

	Master	100 g	10 g	50 g
Dithranol BP	1 g	100 mg	10 mg	50 mg
Zinc and Salicylic Acid Paste BP	999 g	99.9 g	9.99 g	49.95 g

4.   Method of preparation

a.   Solubility where applicable

Not applicable.

b.   Vehicle/diluent

Zinc and Salicylic Acid Paste BP is used as the diluent as per the product formula.

c.   Preservative

There is no preservative included as per the product formula.

d.   Flavouring when appropriate

Pastes are for external use and so no flavouring is required.

The following method would be used to prepare 50 g of Dithranol paste BP 0.1% from the formula above:

1.   Weigh 50 mg Dithranol BP on a Class I or sensitive electronic balance.
2.   Transfer to a glass tile.
3.   Weigh 49.95 g Zinc and Salicylic Acid Paste BP on a Class II or electronic balance.
4.   Transfer the Zinc and Salicylic Acid Paste BP to the glass tile.
5.   Triturate Dithranol BP with the Zinc and Salicylic Acid Paste BP, remembering the principle of 'doubling-up' until a smooth product is formed.
6.   Weigh 40 g of the product and pack into a collapsible tube or amber glass jar. Label and dispense.

Tips

Dithranol BP is extremely irritant and care should be taken when handling. If large quantities are to be made, anecdotal evidence suggests that using Liquid Paraffin BP to dissolve the powder prior to addition to the paste reduces the likelihood of dispersal of the powder when admixing with the paste. If used, the formula would need to be slightly adjusted to allow for the weight of Liquid Paraffin BP used.

5.   Choice of container

A collapsible tube or plain amber jar would be most suitable.

6.   Labelling considerations

a.   Title

The product is official, therefore the following title would be suitable: 'Dithranol Paste BP 0.1%'.

b.   Quantitative particulars

Quantitative particulars are not required as the product is official.

c. Product-specific cautions (or additional labelling requirements)
   'For external use only' will need to be added to the label as the
   product is a paste for external use.
d. Directions to patient – interpretation of Latin abbreviations
   where necessary
   'Apply as directed.'
e. Recommended *British National Formulary* cautions when
   suitable
   The *British National Formulary* (61st edn, p 718) recommends
   the following cautions:
   Label 28 – 'Spread thinly on the affected skin only'.
f. Discard date
   The product is a paste and so will attract a 3-month discard
   date.
g. Sample label (you can assume that the name and address of the
   pharmacy and the words 'Keep out of the reach and sight of
   children' are pre-printed on the label):

Dithranol Paste BP 0.1%	**40 g**
Apply as directed.Spread thinly on the affected skin only	
For external use only	
Do not use after (3 months)	
Miss Amy Smith	Date of dispensing

7. **Advice to patient**
   The patient would be advised to apply the paste thinly as
   directed. In addition, the discard date and the fact that the
   product is for external use only would be highlighted to the
   patient. If further direction for use is requested, the patient
   could be advised to apply the paste once or twice a day as,
   although the concentration of Salicylic Acid BP is quite low
   when compared to preparations used in the treatment of warts
   and hard skin, it will act as a keratolytic and so a maximum
   application frequency of once or twice a day would be
   appropriate.

Self-assessment

1. You are asked to prepare 25 g of a 1 in 5 dilution of Hydrocortisone 1%
   Ointment BP in White Soft Paraffin BP. What quantity of Hydrocortisone 1%
   Ointment BP would be required (the 25 g includes an overage)?
a. 4 g
b. 4.166 g
c. 5 g
d. 5.166 g

2. You are asked to prepare 15 g of an ointment containing 25% Salicylic Acid BP in White Soft Paraffin BP. What quantity of powder is required (the 15 g includes an overage)?
a.   2.5 g
b.   2.75 g
c.   3.75 g
d.   5.75 g

3. How much White Soft Paraffin BP would be used when making the ointment outlined in Question 2?
a.   9.35 g
b.   11.25 g
c.   12.25 g
d.   12.5 g

4. You are asked to prepare 50 g of an ointment containing 0.75% Salicylic Acid BP in White Soft Paraffin BP. What quantity of powder is required (the 50 g includes an overage)?
a.   75 mg
b.   375 mg
c.   3.75 g
d.   7.5 g

5. You are presented with the following prescription: Hydrocortisone BP 2.5 g White Soft Paraffin BP ad 50 g. What is the percentage w/w of Hydrocortisone BP?
a.   1%
b.   2.5%
c.   4.75%
d.   5%

6. You are asked to prepare the following ointment:

Calamine BP	15%
Strong Coal Tar Solution BP	2.5%
Zinc Oxide BP	12.5%
Hydrous Wool Fat BP	25%
White Soft Paraffin BP	ad 100%

How much Strong Coal Tar Solution BP would be required to produce 20 g of this product?
a.   0.5 mL
b.   0.5 g
c.   2.5 mL
d.   0.25 g

7. **How much White Soft Paraffin BP would be in 20 g of the product in Question 6?**
a. 1.8 g
b. 4.5 g
c. 9 g
d. 18 g

8. **The most suitable way to incorporate a coarse insoluble powder into a molten ointment base is by:**
a. trituration
b. fusion
c. levigation
d. titration

9. **A suitable discard date for an extemporaneously prepared ointment would be:**
a. 2 weeks
b. 4 weeks
c. 1 month
d. 3 months

10. **The directions on the prescription for an ointment include the instruction paa. How will this be written on the label?**
a. 'Apply when required.'
b. 'Apply to the vagina.'
c. 'Apply after food.'
d. 'Apply to the affected area.'

11. **Describe the major differences between ointments and creams:**
a. as pharmaceutical formulations
b. as products used by a patient.

Formulation questions

This section contains details of extemporaneous products to be made in the same way as the examples earlier in this chapter. For each example, provide answers using the following sections:

1. **Use of the product**
2. **Is it safe and suitable for the intended purpose?**
3. **Calculation of formula for preparation**
4. **Method of preparation**
    a. Solubility where applicable
    b. Vehicle/diluent
    c. Preservative
    d. Flavouring when appropriate
5. **Choice of container**

6. **Labelling considerations**
a. Title
b. Quantitative particulars
c. Product-specific cautions (or additional labelling requirements)
d. Directions to patient – interpretation of Latin abbreviations where necessary
e. Recommended *British National Formulary* cautions when suitable
f. Discard date
g. Sample label (you can assume that the name and address of the pharmacy and the words 'Keep out of the reach and sight of children' are pre-printed on the label)
7. **Advice to patient.**

12. **You receive a prescription in your pharmacy with the following details:**

Patient:	Mr Amarjit Singh, 6 Summet Drive, Astonbury
Age:	44
Prescription:	Cetrimide Emulsifying Ointment BP
Directions:	Use three times a day
Mitte:	30 g

13. **You receive a prescription in your pharmacy with the following details:**

Patient:	Mrs Helen Preston, 21 Elm Road, Astonbury	
Age:	40	
Prescription:	Salicylic Acid	2%
	Sulphur ppt	3%
	Hydrous oint	qs
Directions:	Apply bd to patches	
Mitte:	15 g	

Overview

Upon completion of this chapter, you should be able to:
- understand how to prepare suppositories and pessaries extemporaneously
- know how to calibrate a suppository/pessary mould
- perform percentage and displacement value calculations during the preparation of suppositories and pessaries
- select an appropriate container in which to package suppositories and pessaries
- prepare an appropriate label for suppositories and pessaries.

Introduction and overview of suppositories and pessaries

Suppositories

British Pharmacopoeia **(BP) definition (suppositories)**

Suppositories are solid, single-dose preparations. The shape, volume and consistency of suppositories are suitable for rectal administration.

They contain one or more active substances dispersed or dissolved in a suitable basis which may be soluble or dispersible in water or may melt at body temperature. Excipients such as diluents, adsorbents, surface-active agents, lubricants, antimicrobial preservatives and colouring matter, authorised by the competent authority, may be added if necessary.

Definition
Pessaries

Common ingredients for inclusion in pessaries for local action include:
- antiseptics
- contraceptive agents
- local anaesthetics
- various therapeutic agents to treat trichomonal, bacterial and monilial infections.

Definition

Suppositories are solid unit dosage forms suitably shaped for insertion into the rectum. The bases used either melt when warmed to body temperature or dissolve or disperse when in contact with mucous secretions. Suppositories may contain medicaments, dissolved or dispersed in the base, which are intended to exert a systemic effect. Alternatively the medicaments or the base itself may be intended to exert a local action. Suppositories are prepared extemporaneously by incorporating the medicaments into the base and the molten mass is then poured at a suitable temperature into moulds and allowed to cool until set.

Definition

Pessaries are a type of suppository intended for vaginal use. The larger-size moulds are usually used in the preparation of pessaries such as 4 g and 8 g moulds. Pessaries are used almost exclusively for local medication, the exception being prostaglandin pessaries that do exert a systemic effect.

KeyPoints

Advantages and disadvantages of suppositories as dosage forms

Advantages
- Can exert local effect on rectal mucosa.
- Used to promote evacuation of bowel.
- Avoid any gastrointestinal irritation.
- Can be used in unconscious patients (e.g. during fitting). Can be used for systemic absorption of drugs and avoid first-pass metabolism.

Disadvantages
- May be unacceptable to certain patients.
- May be difficult to self-administer by arthritic or physically compromised patients.
- Unpredictable and variable absorption *in vivo*.

British Pharmacopoeia (BP) definition (pessaries)

Pessaries are solid, single-dose preparations. They have various shapes, usually ovoid, with a volume and consistency suitable for insertion into the vagina. They contain one or more active substances dispersed or dissolved in a suitable basis that may be soluble or dispersible in water or may melt at body temperature. Excipients such as diluents, adsorbents, surface-active agents, lubricants, antimicrobial preservatives and colouring matter, authorised by the competent authority, may be added, if necessary.

General method

The methods used in the preparation of pessaries are the same as those for suppositories. Within this chapter, points relating to suppositories can also apply to pessaries.

The preparation of suppositories invariably involves some wastage and therefore it is recommended that calculations are made for excess. For example, if you are required to dispense six suppositories, to include a suitable excess, calculate for 10.

Suppository mould calibration

Suppository moulds (Figure 7.1) are calibrated in terms of the weight of Theobroma Oil BP each will contain. Typical sizes are 1 g, 2 g or 4 g. Since the moulds are filled volumetrically, use of a base other than Theobroma Oil BP will require recalibration of the moulds. Many synthetic fats have been formulated to match the specific gravity of Theobroma Oil BP and therefore the mould sizing will be the same and not require recalibration. However, this is not the case for all synthetic bases.

To recalibrate a suppository mould, the compounder needs to prepare a number (e.g. five) of (perfectly formed) suppositories containing only the base. These can then be weighed and the total weight divided by the number of suppositories present to find the mould calibration value.

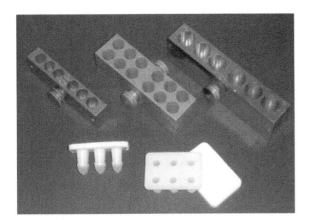

Figure 7.1.
A selection of suppository moulds.

Worked examples
Example 7.1
Calibrate a 1g mould with a synthetic base

1. The synthetic base is melted in an evaporating basin over a water bath until around two-thirds of the base has melted.
2. The evaporating basin is then removed from the heat and stirred, using the residual heat to melt the remaining synthetic base.
3. When the base has cooled to close to its melting point, it is poured into the mould and allowed to overfill slightly.
4. After around 5 minutes, trim the tops and then leave the suppositories to set completely.
5. Weigh all the perfect suppositories (i.e. avoiding any chipped suppositories) and divide the total weight by the number of suppositories weighed. This will give the value that should be used for this particular mould with this base.

General method for suppository preparation (using fatty bases)

1. Most moulds prepare six suppositories, but it is necessary to calculate to include an excess (usually a multiple of 10).
2. Choose a suppository mould to provide the suppositories of the required size (usually a 1g size). Check that the two halves of the mould are matched (numbers are etched on the sides).
3. Check that the mould is clean and assemble the mould but do not overtighten the screw.
4. For some suppository bases it is necessary to lubricate the mould (e.g. use Liquid Paraffin BP), but this is not required when using Hard Fat BP.

5. If the suppository is to contain insoluble, coarse powders, these must be ground down in a glass mortar before incorporation.
6. It is important not to overheat the base, which may change its physical characteristics. Find the melting point of the base and heat it to about 5–10°C less than the melting point. (There should still be some solid base present.) Hold the evaporating basin in the palm of your hand and stir (do not use the thermometer to stir) to complete the melting process.
7. Immiscible liquids and insoluble solids should be incorporated into the fatty base by levigation (wet grinding). The substance should be rubbed into the minimum quantity of molten base on a tile using a spatula. The 'shearing' effect will not be obtained if too much base is used, resulting in a gritty product.
8. The paste obtained in step 7 above should be returned to the evaporating basin with the remainder of the base, stirring constantly.
9. The molten mass should be poured into the mould when it is just about to solidify. (This is usually judged by experience. Look for a slight sheen on the surface of the mass, similar to a skin forming on custard as it cools.)
10. Pour the mass into the mould uniformly in one movement.
11. Allow the mixture to overfill slightly but not to run down the sides of the mould (if this happens, it is likely to be due to the mixture still being too hot).
12. When the suppositories have contracted, but before they have set completely, trim off the excess suppository base. This can easily be achieved by rubbing the flat blade of the spatula over the top of the mould.
13. After further cooling, when the suppositories have set, loosen the screw and tap once sharply on the bench. Remove the suppositories carefully (avoid overhandling or damaging the suppositories with your nails).
14. Pack the required number of suppositories individually in foil and place in an amber wide-necked jar.

See Suppositories video for a demonstration of the preparation and packaging of suppositories.

Alternative methods would be used for the preparation of suppositories (or pessaries) usi3ng non-fatty bases (for example, using glycero-gelatin bases – see Example 7.2).

Example 7.2
The preparation of Glycerol Suppositories BP
You receive a prescription in your pharmacy with the following details:

Patient:	Master Roger Carlson, 21 Hazel Grove, Astonbury
Age:	6
Prescription:	Glycerol Suppositories BP
Directions:	Insert i prn
Mitte:	6

1. **Use of the product**
   Used to treat constipation (*British National Formulary* 61st edn, p 70).
2. **Is it safe and suitable for the intended purpose?**
   This is an official preparation, therefore the formula is safe and suitable for purpose.
3. **Calculation of formula for preparation**
   Prepare 6 child-size Glycerol Suppositories BP.

Product formula
(from the *British Pharmacopoeia* 2007, p 2621)

	Master	100 g	10 g	5 g	25 g
Gelatin BP	14%	14 g	1.4 g	0.7 g	3.5 g
Glycerol BP	70%	70 g	7 g	3.5 g	17.5 g
Purified water	to 100%	16 g	1.6 g	0.8 g	4 g

Calculations
Six suppositories are required; however an overage will be needed to prepare this quantity successfully. Calculations are therefore based on the amounts required to prepare 10 suppositories.

The mass of base that would be needed to prepare 10 child-size (2 g suppositories) is: $10 \times 2 \times 1.2 = 24$ g

For ease of calculation and weighing, sufficient quantities of ingredients to prepare 25 g of base are used.

4. **Method of preparation**
a. Solubility where applicable
   Not applicable.
b. Vehicle/diluent
   Not applicable.
c. Preservative
   There is no preservative included as per the product formula.
d. Flavouring when appropriate Suppositories are for rectal use and so no flavouring is required.

Tips

The choice of suppository mould for a child's glycerol suppository is traditionally 2 g (a 1 g mould is usually used for an infant's glycerol suppository, a 2 g mould for a child's glycerol suppository and a 4 g mould for an adult's glycerol suppository).

The mould will have been calibrated for use with Theobroma Oil BP or Hard Fat BP. The glycero-gelatin base has a greater density, therefore the nominal weight required to fill the moulds will be greater than with Theobroma Oil BP or Hard Fat BP. To calculate correctly the quantities required, the amount that would be required to fill the nominal weight will need to be multiplied by a factor of 1.2.

Tips

Care should be taken when using Arachis Oil BP to ensure that the patient does not have any nut allergies. If this is not known, it would be safest to use Liquid Paraffin BP.

Tips

The weight of the evaporating basin will be needed for the final adjustment of the weight of the product. An excess of water is used to allow for evaporation and, in addition, the use of excess water aids the dissolution of the Gelatin BP.

Tips

Stir gently and not too vigorously to prevent the incorporation of air bubbles.

Tips

This glycero-gelatin base is used as a carrier base for other medicaments in some instances.

Water-soluble thermolabile ingredients are dissolved in a little water before being added to the molten mass.

Insoluble substances are rubbed down on the tile with a little of the Glycerol BP (Glycerol BP and not the base is used as it is difficult to remelt solidified base).

Gelatin BP may be contaminated with pathogenic microorganisms because of its origin. Pharmaceutical-grade gelatin should be pathogen-free, but as an added precaution, when used for pessaries the base may be heat-treated by steam at a temperature of 100°C for 1 hour (this is prior to making up to weight and prior to the addition of any thermolabile ingredients).

Patients with strict religious beliefs and vegetarians may object to the use of animal gelatin (although animal-free gelatin may be available).

Method of preparation of Glycerin Suppositories BP

1. Prepare the mould by lubricating it with either Arachis Oil BP or Liquid Paraffin BP.
2. Weigh approximately 10 g of purified water and place in a previously weighed evaporating basin.
3. Weigh 17.5 g of Glycerol BP on a Class II or electronic balance and place in an evaporating basin.
4. Heat the Glycerol BP over a water bath to 100°C.
5. Weigh 3.5 g Gelatin BP on a Class II or electronic balance.
6. Heat the water to boiling point in a tared beaker.
7. Remove from the heat (to prevent excess evaporation).
8. Add the Gelatin BP powder to the water and stir to dissolve.
9. Add the hot Glycerol BP to the solution and stir until homogeneous.
10. Adjust the weight by adding sufficient hot water or by evaporation of any excess water.
11. Pour the mass into the mould, taking care not to overfill as the base does not contract upon cooling.
12. Leave to cool then remove from the mould, wrap, pack and label.

5. **Choice of container**

Once manufactured, the suppositories should be individually wrapped in foil and placed in an ointment jar. Alternatively, the suppositories could be made in a disposable mould, which can be labelled and dispensed directly to the patient.

6. **Labelling considerations**

a. Title

The product is official, therefore the following title would be suitable: 'Glycerol Suppositories BP (2 g)'.

The '(2 g)' could be added to the title to indicate the size of the Glycerol Suppositories BP as they are available in a number of different sizes.

b. Quantitative particulars
    Quantitative particulars are not required as the product is official.
c. Product-specific cautions (or additional labelling requirements)
    'For rectal use only' will need to be added to the label as the products are suppositories for rectal use.
d. Directions to patient – interpretation of Latin abbreviations where necessary
    'Insert ONE into the rectum when required.'
e. Recommended *British National Formulary* cautions when suitable
    Not applicable.
f. Discard date
    The products are suppositories and so will attract a 3-month discard date.
g. Sample label (you can assume that the name and address of the pharmacy and the words 'Keep out of the reach and sight of children' are pre-printed on the label):

Glycerol Suppositories BP (2 g)	6
Insert ONE into the rectum when required.	
For rectal use only	
Do not use after (3 months)	
Master Roger Carlson	Date of dispensing

7. **Advice to patient**
    The parent/guardian would be advised to insert one suppository into the rectum when required. In addition, the discard date and the fact that the product is for rectal use only would be highlighted to the patient.

Example 7.3
The preparation of Ichthammol BP Pessaries 5%

You receive a prescription in your pharmacy with the following details:

Patient:	Miss Jenny Tibbs, 64 Farmington Avenue, Astonbury
Age:	26
Prescription:	Ichthammol Pessaries 5%
Directions:	Insert i prn
Mitte:	3

1. **Use of the product**
   Used to treat cervicitis and vaginitis (*Martindale* 26th edn, p 565).
2. **Is it safe and suitable for the intended purpose?**
   The formula is the same as Ichthammol Pessaries BPC (*British Pharmaceutical Codex* 1968, p 1222).
3. **Calculation of formula for preparation**
   Prepare 3 × 8 g Ichthammol BP Pessaries 5%.

	Master	100 g	10 g	50 g
Ichthammol BP	5 %	5 g	0.5 g	2.5 g
Glycero-gelatin Base BP	95 %	95 g	9.5 g	47.5 g

Tips

The Glycero-gelatin Base BP has a greater density, therefore the nominal weight required to fill the moulds will be greater than with the other two bases (Theobroma Oil BP and Hard Fat BP). To calculate correctly the quantities required, the amount that would be required to fill the nominal weight will need to multiplied by a factor of 1.2.

Calculations

Three pessaries are required; however, an overage will be needed to prepare this quantity successfully. Calculations are therefore based on the amounts required to prepare five pessaries.

The mass of base that would be needed to prepare five pessaries is:

$5 \times 8 \times 1.2 = 48\,g$

For ease of calculation and weighing, sufficient quantities of ingredients to prepare 50 g of base are used.

	Master	100 g	50 g
Gelatin BP	14%	14 g	7 g
Glycerol BP	70%	70 g	35 g
Purified water	to 100%	16 g	8 g

Tips

The *British Pharmacopoeia* states that purified gelatin can be produced by two different methods.
1. partial acid hydrolysis (type A (anionic) gelatin)
2. partial alkaline hydrolysis (type B (cationic) gelatin).

Because the active ingredient to be incorporated in these pessaries is Ichthammol BP, type B should be used to avoid incompatibilities.

4. **Method of preparation**
a. Solubility where applicable
   Not applicable.
b. Vehicle/diluent
   Purified water is used as per the product formula.
c. Preservative
   There is no preservative included as per the product formula.
d. Flavouring when appropriate
   Pessaries are for vaginal use and so no flavouring is required.

Method of preparation of Ichthammol BP Pessaries 5%

1. Prepare the mould by lubricating it with either Arachis Oil BP or Liquid Paraffin BP.
2. Prepare the base as in Example 7.2 above.
3. Weigh 2.5 g of Ichthammol BP on a Class II or electronic balance.
4. Weigh 47.5 g of base
5. Remove the base from the heat and add the Ichthammol BP.
6. Stir until homogeneous and then pour into the lubricated mould.
7. Leave the pessaries to cool, remove from the mould, wrap, pack and label.

5. **Choice of container**
   Once manufactured, the pessaries should be individually wrapped in foil and placed in an ointment jar. Alternatively, the pessaries could be made in a disposable mould, which can be labelled and dispensed directly to the patient.

6. **Labelling considerations**

a. Title
   The product is unofficial, therefore the following title would be suitable: 'Ichthammol Pessaries 5%'.

b. Quantitative particulars
   Quantitative particulars are required as the product is unofficial. As the products are pessaries, manufactured by percentages, the quantitative particulars will be expressed per pessary in percentages:

Each pessary contains:

Ichthammol BP	5%
Glycero-gelatin Base BP	95%

c. Product-specific cautions (or additional labelling requirements)
   'For vaginal use only' will need to be added to the label as the products are pessaries for vaginal use.

d. Directions to patient – interpretation of Latin abbreviations where necessary
   'Insert ONE into the vagina when required.'

e. Recommended *British National Formulary* cautions when suitable
   Not applicable.

Tips

Care should be taken when using Arachis Oil BP to ensure that the patient does not have any nut allergies. If this is not known, it would be safest to use Liquid Paraffin BP.

Tips

Because the base was prepared in a tared container, it is easier to remove any excess base rather than try to weigh the amount of base required and transfer to another vessel.
   Note: 2.5 g of the base needs to be replaced by Ichthammol BP.
   This amount of base needs to be removed. It would not be correct to adjust the base weight by evaporation as this would mean that the Ichthammol BP was replacing water, not base.

f. Discard date

The products are pessaries and so will attract a 3-month discard date.

g. Sample label (you can assume that the name and address of the pharmacy and the words 'Keep out of the reach and sight of children' are pre-printed on the label):

Ichthammol Pessaries 5%		3
	Insert ONE into the vagina when required.	
	For vaginal use only	
	Do not use after (3 months)	
Each suppository contains:		
Ichthammol BP	5%	
Glycero-gelatin Base BP	95%	
Miss Jenny Tibbs		Date of dispensing

7. Advice to patient

The patient would be advised to insert one pessary into the vagina when required. In addition, the discard date and the fact that the product is for vaginal use only would be highlighted to the patient.

Example 7.4
The preparation of Compound Bismuth Subgallate Suppositories BP

You receive a prescription in your pharmacy with the following details.

Patient:	Mr Arnold Beach, 2 Springfield Road, Astonbury
Age:	38
Prescription:	Compound Bismuth Subgallate Suppositories BP
Directions:	Insert i prn
Mitte:	6

1. Use of the product

Used as an astringent to treat haemorrhoids (*British National Formulary* 61st edn, p 75).

2. Is it safe and suitable for the intended purpose?

This is an official preparation therefore the formula is safe and suitable for purpose.

3. Calculation of formula for preparation

Prepare 6 suppositories.

Product formula
(from the *British Pharmacopoeia* 1980, p 723)

	1 suppository	10 suppositories
Bismuth Subgallate BP	200 mg	2 g
Resorcinol BP	60 mg	600 mg
Zinc Oxide BP	120 mg	1.2 g
Castor Oil BP	60 mg	600 mg
Base	qs	qs

Calculations

Prepare for 10 suppositories to allow for losses during preparation.

Displacement values

Bismuth Subgallate BP	2.7
Resorcinol BP	1.5
Zinc Oxide BP	4.7
Castor Oil BP	1.0

Bismuth Subgallate BP displaces 2 ÷ 2.7 g Hard Fat BP	= 0.74 g
Resorcinol BP displaces 0.6 ÷ 1.5 g Hard Fat BP	= 0.4 g
Zinc Oxide BP displaces 1.2 ÷ 4.7 g Hard Fat BP	= 0.26 g
Castor Oil BP displaces 0.6 ÷ 1 g Hard Fat BP	= 0.6 g
The amount of base required	= (10 × 1 g) − (0.74 + 0.4 + 0.26 + 0.6)
	= 10 − 2.00
	= 8.00 g

Tips

Hard Fat BP is a suitable base to use. The quantity to be used must be calculated using displacement values. A nominal 1 g mould is used.

Working formula

	10 suppositories
Bismuth Subgallate BP	2 g
Resorcinol BP	600 mg
Zinc Oxide BP	1.2 g
Castor Oil BP	600 mg
Hard Fat BP	8 g

4. **Method of preparation**
a. Solubility where applicable
   Not applicable.
b. Vehicle/diluent
   Hard Fat BP is being used as the base for this preparation.
c. Preservative
   There is no preservative included as per the product formula.
d. Flavouring when appropriate
   Suppositories are for rectal use and so no flavouring is required.

Method of preparation

Noting that the melting point of Hard Fat BP is 30–45°C (*Martindale* 35th edn, p 1847).

1. Weigh 8 g of Hard Fat BP on a Class II or electronic balance.
2. Transfer to an evaporating basin and melt over a water bath.
3. Weigh 2 g of Bismuth Subgallate BP on a Class II or electronic balance.
4. Weigh 600 mg Resorcinol BP on a Class II or electronic balance.
5. Weigh 1.2 g Zinc Oxide BP on a Class II or electronic balance.
6. Weigh 600 mg Castor Oil BP on a Class II or electronic balance.
7. Mix the powders together in a mortar using the 'doubling-up' technique and transfer to a warmed tile.
8. Levigate the powders with the Castor Oil BP and a little molten base.
9. Return the resultant mix to the molten mass and stir.
10. Stir until almost set, then transfer to a clean, dry, matched suppository mould and allow to set.
11. Trim the tops and remove from the mould.
12. Wrap individually in foil and transfer to an amber glass jar and label.

Tips

Any liquid ingredients to be added to suppositories must be weighed, *not* measured.

5. **Choice of container**

   Once manufactured, the suppositories should be individually wrapped in foil and placed in an ointment jar. Alternatively, the suppositories could be made in a disposable mould, which can be labelled and dispensed directly to the patient.

6. **Labelling considerations**

a. Title

   The product is official, therefore the following title would be suitable: 'Compound Bismuth Subgallate Suppositories BP'.

b. **Quantitative particulars**

   Quantitative particulars are not required as the product is official.

c. Product-specific cautions (or additional labelling requirements)

   'For rectal use only' will need to be added to the label as the products are suppositories for rectal use.

d. Directions to patient – interpretation of Latin abbreviations where necessary

   'Insert ONE into the rectum when required.'

e. Recommended *British National Formulary* cautions when suitable

   Not applicable.

f. Discard date

   The products are suppositories and so will attract a 3-month discard date.

g. Sample label (you can assume that the name and address of the pharmacy and the words 'Keep out of the reach and sight of children' are pre-printed on the label):

Compound Bismuth Subgallate Suppositories BP	**3**
Insert ONE into the rectum when required.	
For rectal use only	
Do not use after (3 months)	
Mr Arnold Beach	Date of dispensing

7. **Advice to patient**
The patient would be advised to insert one suppository into the rectum when required after defecation. In addition, the discard date and the fact that the product is for rectal use only would be highlighted to the patient.

Self-assessment

1. **Which of the following cautions is required on the label of all suppositories?**
a. 'Warm to body temperature before use.'
b. 'For rectal use only.'
c. 'For external use only.'
d. 'Store in a cool place.'

2. **Suppositories will normally attract a discard date of:**
a. 2 weeks
b. 4 weeks
c. 1 month
d. 3 months

3. **You are asked to prepare six 1 g suppositories each containing 200 mg of Zinc Oxide BP (displacement value = 4.7). How much Hard Fat BP base would be contained in each suppository?**
a. 800 mg
b. 857.5 mg
c. 900 mg
d. 957.5 mg

4. **You are asked to prepare 10 × 1 g suppositories, each containing Phenobarbital BP 60 mg (displacement value = 1.1). How much Hard Fat BP base would be contained in each suppository?**
a. 905 mg
b. 940 mg

c.   945 mg
d.   998.9 mg

5.   **Which of the following suppositories exerts a systemic effect?**
a.   Anusol
b.   Bisacodyl
c.   Proctosedyl
d.   Stemetil

6.   **Which of the following statements is true?**
a.   A suppository base should melt at just above 40°C.
b.   A suppository base should melt at just above 37°C.
c.   A suppository base should melt at just below 37°C.
d.   A suppository base should melt at just below 25°C.

7.   **List the properties of an ideal suppository base.**

Formulation questions

This section contains details of extemporaneous products to be made in the same way as the examples earlier in this chapter. For each example, provide answers using the following sections:

1.   **Use of the product**
2.   **Is it safe and suitable for the intended purpose?**
3.   **Calculation of formula for preparation**
4.   **Method of preparation**
a.   Solubility where applicable
b.   Vehicle/diluent
c.   Preservative
d.   Flavouring when appropriate
5.   **Choice of container**
6.   **Labelling considerations**
a.   Title
b.   Quantitative particulars
c.   Product-specific cautions (or additional labelling requirements)
d.   Directions to patient – interpretation of Latin abbreviations where necessary
e.   Recommended *British National Formulary* cautions when suitable
f.   Discard date
g.   Sample label (you can assume that the name and address of the pharmacy and the words 'Keep out of the reach and sight of children' are pre-printed on the label)
7.   **Advice to patient**

8. You receive a prescription in your pharmacy with the following details:

Patient:	Miss Jessica Ramsden, 12 Bishop Road, Astonbury
Age:	4
Prescription:	Supps Paracetamol 180 mg
Directions:	1 pr qds prn
Mitte:	4

9. You receive a prescription in your pharmacy with the following details:

Patient:	Master Samuel Bridges, 5 Meadow Way, Astonbury
Age:	4
Prescription:	Suppositories Metronidazole 170 mg
Directions:	1 pr tds
Mitte:	2/7

Overview

Upon completion of this chapter, you should be able to:

- prepare bulk powders for external use (dusting powders), bulk oral powders, individual unit dose powders and unit dose capsules
- perform both single and double dilution calculations
- select an appropriate container in which to package powders and capsules
- prepare an appropriate label for powders and capsules.

Introduction and overview of powders and capsules

This section will include solid preparations intended for both internal and external use. The following types of preparation will be considered:

- bulk powders for external use – termed dusting powders
- bulk oral powders
- individual unit dose powders
- unit dose capsules.

General method

General method for preparing dusting powders

The method for mixing powders in the formulation of a dusting powder is the standard 'doubling-up' technique.

'Doubling-up' technique

1. Weigh the powder present in the smallest volume (powder A) and place in the mortar.
2. Weigh the powder present in the next largest volume (powder B) and place on labelled weighing paper.
3. Add approximately the same amount of powder B as powder A in the mortar.
4. Mix well with pestle.

KeyPoints

Advantages and disadvantages of dusting powders as dosage forms

Advantages
- Easy to apply
- Pleasant to use
- Absorb skin moisture

Disadvantages
- May block pores causing irritation
- Possibility of contamination
- Light fluffy powders may be inhaled by infants leading to breathing difficulties
- Not suitable for application to broken skin

KeyPoints

Advantages and disadvantages of bulk oral powders as dosage forms

Advantages
- May be more stable than liquid equivalent
- Administered with relative ease
- Absorption quicker than capsules or tablets

Disadvantages
- Variable dose accuracy
- Bulky and inconvenient to carry
- Difficult to mask unpleasant tastes

KeyPoints

Advantages and disadvantages of suspensions as dosage forms

Advantages
- More stable than liquid dosage forms
- Accurate dosing
- Easy to administer
- Small particle size of drug
- Acceptable to patients

Disadvantages
- May be difficult to swallow
- Hard to mask unpleasant flavours

KeyPoints

Advantages and disadvantages of unit dose capsules as dosage forms

Advantages
- More stable than liquid dosage forms
- Accurate dosing
- Easy to administer
- Unpleasant tastes easily masked
- Release characteristics can be controlled
- Can be made light-resistant
- Small particle size of drug
- Acceptable to patients

Disadvantages
- May be difficult to swallow
- Unsuitable for very small children
- Possible patient objections to the use of animal gelatin

5. Continue adding an amount of powder B that is approximately the same as that in the mortar and mix with the pestle, i.e. doubling the amount of powder in the mortar at each addition.

6. If further powders are to be added, add these in increasing order of volume as in steps 3, 4 and 5 above.

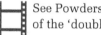 See Powders video for a demonstration of the 'doubling-up' technique.

General method for producing unit dose powders

1. Remember, for ease of handling the minimum weight of powder in a unit dose paper is 200 mg.

2. Calculate to make an excess of the number of powders requested.

3. Determine whether a single or double dilution of the active ingredient is required.

4. Mix the active ingredient and the diluent (Lactose BP unless there is a reason not to use it, for example, if the patient is intolerant to lactose or due to instability of the ingredients) in a mortar using the 'doubling-up' technique.

5. Work on a clean dry glass tile. Select a suitable size of paper (e.g. 10 × 10 cm), turn in one edge and fold down approximately 1.25 cm. Repeat for the required number of powders.

6. Place the paper on the glass tile, with the folded edge away from the compounder, and each edge slightly overlapping, next to the balance pan to be used for weighing.

7. Weigh out the individual powder from the bulk powder, and transfer to the centre of the paper (if placed too near the fold, the powder will fall out during opening).

8. Fold the bottom of the powder paper up to, and underneath, the flap folded originally.

9. Fold down the top of the paper until it covers about two-thirds of the width of the paper. This top edge of this fold should help to hold the contents of the paper in the centre of the paper.

Figure 8.1.
A powder trestle.

10. Fold the two ends under, so that the loose ends slightly overlap, and then tuck one flap inside the other.
11. Wrap each powder in turn, making sure they are all the same size.
12. Stack the powders in pairs, flap to flap.
13. Tie together with a rubber band (not too tightly).
14. Place in a rigid cardboard box.
15. The label should be placed on the outer pack such that when the patient opens the box, the label is not destroyed.

See Powders video for a demonstration of the preparation of a unit dose powder.

Tips

Historically, a powder trestle (Figure 8.1) was used to assist the compounder with step 10 above. The paper containing the powder was placed centrally on the top of the trestle and then the sides were bent underneath using the edges of the trestle to form neat creases. The use of a powder trestle ensured that all the powders would be of a uniform size.

General method of preparation of capsules

1. Choose an appropriate size capsule for the powder bulk. Normally a size 3 capsule would be chosen and so work on the basis of filling each capsule with 200 mg of powder.
2. Calculate quantities required and make an excess as with the manufacture of individual unit dose powders.
3. Mix using the 'doubling-up' technique.
4. Handle the capsules as little as possible as powder fill weights will be inaccurate as a result of contamination with grease and moisture. This is also important for reasons of hygiene. Fill powder into the longer half of the capsule.
5. There are at least three methods of filling capsules manually. Always work on a clean tile: remember that these capsules are to be swallowed by a patient.
6. Ensure the capsule outer surface is powder-free. Check the weight of the filled capsule. Remember to tare with an empty capsule of the same size so you are only weighing the contents

of the capsule (and not including the weight of the capsule itself).

See Powders video for a demonstration of the preparation of a capsule.

Worked examples

Example 8.1
The preparation of Zinc, Starch and Talc Dusting Powder BPC

You receive a prescription in your pharmacy with the following details:

Patient:	Mr James Davenport, 34 Smithfield Avenue, Astonbury
Age:	56
Prescription:	Zinc, Starch and Talc Dusting Powder BPC
Directions:	Use mdu
Mitte:	100 g

1. **Use of the product**
   The product is a dusting powder (*Martindale* 33rd edn, p 1124).
2. **Is it safe and suitable for the intended purpose?**
   This is an official formula for a dusting powder, therefore the formula and frequency of application are safe.
3. **Calculation of formula for preparation**
   Prepare 100 g of Zinc Starch and Talc Dusting Powder BPC.

Product formula
(from the *British Pharmaceutical Codex* 1973, p 664)

	Master	100 g
Zinc Oxide BP	250 g	25 g
Starch BP	250 g	25 g
Purified Talc BP	500 g	50 g

4. **Method of preparation**
a. Solubility where applicable
   Not applicable.
b. Vehicle/diluent
   Not applicable.
c. Preservative
   No preservative is included in the preparation.
d. Flavouring when appropriate
   Dusting powders are for external use and so no flavouring is required.

Method for preparing Zinc Starch and Talc Dusting Powder BPC using the above formula

1. Weigh 25 g Zinc Oxide BP using a Class II or electronic balance.
2. Weigh 25 g Starch BP using a Class II or electronic balance.
3. Weigh 50 g Purified Talc BP using a Class II or electronic balance.
4. Transfer the Starch BP to a porcelain mortar.
5. Add the Zinc Oxide BP to the Starch BP in the mortar and mix using a pestle.
6. Add the Purified Talc BP to the powders in the mortar and continue mixing.
7. Transfer the mixed powder to a powder shaker container (Figure 8.2) or an amber glass jar.
8. Label and dispense to the patient.

Tips

The powders are admixed in order of volume, remembering the 'doubling-up' technique.

5. Choice of container

If available, a powder shaker with a sifter top would be the container of choice. Alternatively, an ointment jar could be used.

6. Labelling considerations

a. Title

The product is official, therefore the following title would be suitable: 'Zinc, Starch and Talc Dusting Powder BPC'.

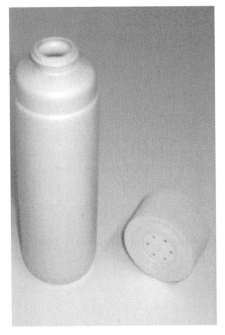

Figure 8.2.
A powder shaker.

b. Quantitative particulars
Quantitative particulars are not required as the product is official.

c. Product-specific cautions (or additional labelling requirements)
'For external use only' will need to be added to the label as the product is a dusting powder for external use.
'Store in a dry place' will need to be added to the label as the product is a dusting powder.
'Not to be applied to open wounds or raw weeping surfaces' will need to be added to the label as the product is a dusting powder.

d. Directions to patient – interpretation of Latin abbreviations where necessary
'Use as directed.'

e. Recommended *British National Formulary* cautions when suitable
Not applicable.

f. Discard date
The product is a dusting powder and so will attract a 3-month discard date.

g. Sample label (you can assume that the name and address of the pharmacy and the words 'Keep out of the reach and sight of children' are pre-printed on the label):

Zinc, Starch and Talc Dusting Powder BPC	**100 g**
Use as directed.	
For external use only	
Store in a dry place	
Not to be applied to open wounds or raw weeping surfaces	
Do not use after (3 months)	
Date of dispensing	Mr James Davenport

7. **Advice to patient**
The patient would be advised to use the product as a dusting powder as directed. In addition, the discard date and the fact that the product is for external use only and should be stored in a dry place would be highlighted to the patient.

Example 8.2

The preparation of Compound Magnesium Trisilicate Oral Powder BP

You receive a prescription in your pharmacy with the following details:

Patient:	Mrs Sally Frost, 345a Longfield Avenue, Astonbury
Age:	56
Prescription:	Compound Magnesium Trisilicate Oral Powder BP
Directions:	1 g tds prn
Mitte:	100 g

1. **Use of the product**
   Used as an antacid (see references below for individual ingredients).
2. **Is it safe and suitable for the intended purpose?**
   This is an official formula for a bulk oral powder, therefore the formula is safe. The dose of 1 g three times a day falls within the ranges for the individual ingredients (*Martindale* 35th edn, pp 1572, 1545, 1571 and 1508).
3. **Calculation of formula for preparation**
   Prepare 100 g of Compound Magnesium Trisilicate Oral Powder BP.

**Product formula
(from the *British Pharmacopoeia* 2007, p 2726)**

	Master	100 g
Magnesium Trisilicate BP	250 g	25 g
Chalk (Powdered) BP	250 g	25 g
Heavy Magnesium Carbonate BP	250 g	25 g
Sodium Bicarbonate BP	250 g	25 g

4. **Method of preparation**
a. Solubility where applicable
   Not applicable.
b. Vehicle/diluent
   Not applicable.
c. Preservative
   No preservative is included in the preparation as per the product formula.
d. Flavouring when appropriate
   No flavouring is included in the preparation as per the product formula.

Method for preparing Compound Magnesium Trisilicate Oral Powder BP using the above formula
1. Weigh 25 g Magnesium Trisilicate BP using a Class II or electronic balance.
2. Weigh 25 g Chalk BP using a Class II or electronic balance.
3. Weigh 25 g Heavy Magnesium Carbonate BP using a Class II or electronic balance.

Tips

The suggested order is Sodium Bicarbonate BP first, then add Chalk BP, Heavy Magnesium Carbonate BP and finally Magnesium Trisilicate BP. The Sodium Bicarbonate BP is noticeably the smallest volume and also the most likely to suffer clumping and be lumpy in appearance.

4. Weigh 25 g Sodium Bicarbonate BP using a Class II or electronic balance.

5. Mix the powders in a porcelain mortar in order of bulk volume.

6. Transfer to an amber glass jar, label and dispense to the patient.

5. **Choice of container**

Once prepared, the powder should be packaged in an ointment jar.

6. **Labelling considerations**

a. Title

The product is official, therefore the following title would be suitable: 'Compound Magnesium Trisilicate Oral Powder BP'.

b. Quantitative particulars

Quantitative particulars are not required as the product is official.

c. Product-specific cautions (or additional labelling requirements)

'Store in a dry place' will need to be added to the label as the product is a bulk oral powder.

d. Directions to patient – interpretation of Latin abbreviations where necessary

'1 g to be mixed with a small amount of water and taken THREE times a day when required.'

e. Recommended *British National Formulary* cautions when suitable

Not applicable.

f. Discard date

The product is a bulk oral powder and so will attract a 3-month discard date.

g. Sample label (you can assume that the name and address of the pharmacy and the words 'Keep out of the reach and sight of children' are pre-printed on the label):

Compound Magnesium Trisilicate Oral Powder BP	**100 g**
1 g to be mixed with a small amount of water and taken THREE times a day when required. Store in a dry place Do not use after (3 months)	
Mrs Sally Frost	Date of dispensing

7. **Advice to patient**

The patient would be advised to take 1 g mixed with a small amount of water three times a day when required. She would be advised that the easiest way to achieve this at home would be to use a level teaspoon of powder, which is approximately equivalent to 1 g. In addition, the discard date and the fact that the product is to be stored in a dry place would be highlighted to the patient.

Example 8.3
The preparation of five individual dose powders of Codeine Phosphate BP 10 mg

You receive a prescription in your pharmacy with the following details:

Patient:	Mr Tommy Jacks, 13 Albert Drive, Astonbury
Age:	76
Prescription:	Codeine Phosphate 10 mg powders
Directions:	i prn
Mitte:	6

1. **Use of the product**
    Codeine Phosphate BP is used to treat mild to moderate pain and diarrhoea and as a cough suppressant (*British National Formulary* 61st edn, p 264).
2. **Is it safe and suitable for the intended purpose?**
    The usual adult dose to treat pain is 30–60 mg every 4 hours when necessary to a maximum of 240 mg daily (*British National Formulary* 61st edn, p 264). However, consultation with the prescriber indicates that the patient is frail and requires a lower than normal dose of codeine to achieve pain control. The preparation is therefore safe and suitable for the intended purpose.
3. **Calculation of formula for preparation**

Product formula

	1 powder	10 powders
Codeine Phosphate BP	10 mg	100 mg
Lactose BP	to 200 mg	to 2000 mg

Calculation

The quantity of Codeine Phosphate BP required for the 10 powders is 100 mg, which is below the minimum weighable quantity for a Class II balance. Therefore it is recommended to follow the double (serial) dilution process.

A concentrated powder where every 200 mg of this concentrate (mix X) contains 100 mg Codeine Phosphate BP (mix X contains 100 mg/200 mg) needs to be prepared.

As 100 mg cannot be accurately weighed, the quantities in mix X need to be adjusted. To keep mix X the same concentration, both parts of the concentration ratio must be multiplied by the same factors:

2×100 mg = 200 mg

2×200 mg = 400 mg

> **Tips**
>
> The final weight of individual powders that we recommend for ease of calculation and administration is 200 mg. An excess is made to allow for losses during preparation.

Therefore mix X must have a concentration 200 mg/400 mg (200 mg Codeine Phosphate BP per 400 mg of mix X).

As we must have exact weights, the quantities for mix X are:

Codeine Phosphate BP	200 mg
Lactose BP	200 mg (i.e. to 400 mg)

Therefore the final formula for preparation for the 10 powders, mix Y, will be:

Mix X	200 mg (containing 100 mg Codeine Phosphate BP)
Lactose BP	to 2000 mg (1800 mg)

4. **Method of preparation**

a. Solubility where applicable
   Not applicable.
b. Vehicle/diluent
   Lactose BP is used as a diluent (unless the patient is lactose-intolerant).
c. Preservative
   No preservative is included in the preparation.
d. Flavouring when appropriate
   Oral powders are swallowed with a draught of water and, as such, do not require flavouring.

Method for preparing Codeine Phosphate 10 mg unit dose powders using the above formula

1. Weigh 200 mg Codeine Phosphate BP using a Class II or electronic balance.
2. Transfer to a porcelain mortar.
3. Weigh 200 mg Lactose BP using a Class II or electronic balance.
4. Add the Lactose BP to the Codeine Phosphate BP in the mortar using the 'doubling-up' technique.
5. This is mix X.
6. Weigh 200 mg mix X using a Class II or electronic balance and transfer to a clean dry mortar.
7. Weigh 1800 mg Lactose BP using a Class II or electronic balance.
8. Add the Lactose BP to the mix X in the mortar using the 'doubling-up' technique.
9. This is mix Y.
10. Weigh 200 mg aliquots of the mix Y and wrap as individual dose powders.
11. Pack the powders flap to flap and enclose with a rubber band.
12. Pack into a cardboard box and label.

5. **Choice of container**

   Once manufactured, the powders should be packaged flap to flap and enclosed with a rubber band. They can then be placed in a cardboard carton.

6. **Labelling considerations**

a. Title

   The product is unofficial, therefore the following title would be suitable: 'Codeine Phosphate 10 mg powders'.

b. Quantitative particulars

   Quantitative particulars are required as the product is unofficial. As the products are powders for internal administration, the quantitative particulars will be expressed per powder:

   Each powder contains:

Codeine Phosphate BP	10 mg
Lactose BP	190 mg

c. Product-specific cautions (or additional labelling requirements)

   'Store in a dry place' will need to be added to the label as the products are individual dose powders.

d. Directions to patient – interpretation of Latin abbreviations where necessary

   'The contents of ONE powder to be taken when required.'

e. Recommended *British National Formulary* cautions when suitable.

   The *British National Formulary* (61st edn, p 265) recommends the following caution:

   Label 2 – 'Warning: This medicine may make you sleepy. If this happens, do not drive or use tools or machines. Do not drink alcohol.'

f. Discard date

   The products are individual unit dose powders and so will attract a 3-month discard date.

g. Sample label (you can assume that the name and address of the pharmacy and the words 'Keep out of the reach and sight of children' are pre-printed on the label):

Codeine Phosphate 10 mg powders   **6**

The contents of ONE powder to be taken when required.
Store in a dry place
Do not use after (3 months)

Warning: This medicine may make you sleepy. If this happens, do not drive or use tools or machines.
Do not drink alcohol.

Each powder contains:

Codeine Phosphate BP	10 mg
Lactose BP	190 mg
Mr Tommy Jacks	Date of dispensing

7. **Advice to patient**
   The patient would be advised to take the contents of one powder with a glass of water when required. In addition, the discard date, the need to store the product in a dry place and the additional *British National Formulary* warning would be highlighted.

Self-assessment

1. **Which of the following directions should always be included on the label of single unit dose powders?**
   a. 'Store in a dry place.'
   b. 'Store below 12°C.'
   c. 'Do not swallow in large amounts.'
   d. 'Not to be taken.'

2. **Which of the following reference books contains an appendix advising on cautionary and advisory labelling requirements?**
   a. *Martindale. The Extra Pharmacopoeia.*
   b. *(British) Pharmaceutical Codex.*
   c. *British National Formulary.*
   d. *British Pharmacopoeia*, vol. 1.

3. **What labelling instruction is needed on single dose powders but not on capsules?**
   a. Quantitative particulars
   b. Storage conditions
   c. Patient's name
   d. Date of dispensing

4. **How many capsules would you provide on a prescription with instructions '1 tds 5/7'?**
   a. 3
   b. 5
   c. 15
   d. 21

5. **How many capsules would you supply if the directions on the prescription read '2 qqh 1/52'?**
   a. 8
   b. 12
   c. 56
   d. 84

6. **What quantity of capsules would you dispense if a prescription gave the directions '1 tds increasing to 1 qds after a fortnight. Mitte 56/7'?**
   a. 42
   b. 98

c. 168
d. 210

7. **What does trituration mean?**
a. To wet grind
b. To titrate
c. To mix
d. To levigate

8. **Which size capsule would be the capsule of choice when dispensing 200 mg of powder?**
a. Size 0
b. Size 1
c. Size 2
d. Size 3

9. **Which of the following statements is false?**
a. Bulk oral powders are suitable for the administration of relatively non-potent medicaments.
b. Dusting powders are used to lubricate the skin.
c. Dusting powders must be labelled 'Not to be taken'.
d. Starch is a suitable diluent for single dose powders.

10. **Which of the following weights of active ingredient per powder or capsule would necessitate the use of the double dilution technique (assuming you are making a total mix for 10 capsules using a Class II balance)?**
a. 10 mg
b. 20 mg
c. 50 mg
d. 100 mg

11. **You are requested to prepare three powders each containing Ephedrine BP 3 mg. Assuming you make 10 powders, each weighing 200 mg, to allow a suitable excess, which of the following would be the most suitable formula for your concentrated powder mix X?**
a. Ephedrine BP 150 mg Lactose BP 1000 mg
b. Ephedrine BP 150 mg Lactose BP 850 mg
c. Ephedrine BP 200 mg Lactose BP 2000 mg
d. Ephedrine BP 200 mg Lactose BP 1800 mg

12. **You prepare five capsules, each containing 20 mg of Verapamil BP per 200 mg. How would the quantitative particulars be expressed on the label?**
a. Verapamil BP 20 mg Lactose BP 200 mg
b. Verapamil BP 100 mg Lactose BP 1000 mg
c. Verapamil BP 200 mg Lactose BP 1800 mg
d. Verapamil BP 20 mg Lactose BP 180 mg

Formulation questions

This section contains details of extemporaneous products to be made in the same way as the examples earlier in this chapter. For each example, provide answers using the following sections:

1. Use of the product
2. Is it safe and suitable for the intended purpose?
3. Calculation of formula for preparation
4. Method of preparation
   a. Solubility where applicable
   b. Vehicle/diluent
   c. Preservative
   d. Flavouring when appropriate
5. Choice of container
6. Labelling considerations
   a. Title
   b. Quantitative particulars
   c. Product-specific cautions (or additional labelling requirements)
   d. Directions to patient – interpretation of Latin abbreviations where necessary
   e. Recommended *British National Formulary* cautions when suitable
   f. Discard date
   g. Sample label (you can assume that the name and address of the pharmacy and the words 'Keep out of the reach and sight of children' are pre-printed on the label)
7. Advice to patient

13. You receive a prescription in your pharmacy with the following details:

Patient:	Mrs Grace Browning, 47 Bridgeport Street, Astonbury
Age:	34
Prescription:	Bendroflumethiazide 2.5 mg powders
Directions:	1 mane
Mitte:	5/7

14. You receive a prescription in your pharmacy with the following details:

Patient:	Miss Jessica Felsham, 4 Farnham Place, Astonbury
Age:	10
Prescription:	Imipramine Caps 50 mg
Directions:	1 nocte
Mitte:	6

Answers to self-assessment

Chapter 2: Solutions

1. The final solution has a required strength of 15 mg in every 10 mL. Therefore, we require 150 mg (15 × 10) in every 100 mL (10 × 10). Therefore, we require 150 × 5 = 750 mg in 500 mL (100 × 5). **Answer:** c (750 mg)

2. 30 mg of the ingredient is dissolved in 1.5 mL. Therefore, 30 ÷ 1.5 = 20 mg in every 1 mL. **Answer:** c (20 mg/mL)

3. The stock solution contains 600 micrograms in every 1 mL. The patient requires a dose of 1 mg.
   1 ÷ 0.6 = 1.667 mL. As the syringe is graduated in 0.1 mL, the required dose is 1.7 mL. **Answer:** c (1.7 mL)

4. The stock solution contains 0.25 mg per 1 mL. 0.25 × 0.2 = 0.05 mg = 50 micrograms. **Answer:** a (50 micrograms)

5. The required solution contains 6 g/L. Therefore, there are 6 ÷ 1000 = 6 mg/mL. 6 × 150 = 900 mg = 0.9 g. **Answer:** d (0.9 g)

6. The stock solution contains 25 mg/5 mL. Therefore, the stock solution contains 5 mg per 1 mL. **Answer:** 1 mL

7. The stock solution contains 50 mg/5 mL. Therefore, for a dose of 37.5 mg, we require (37.5 ÷ 50) × 5 = 3.75 mL. **Answer:** 3.75 mL

8. The stock solution contains 200 mg/5 mL. Therefore, for a dose of 125 mg, we require (125 ÷ 200) × 5 = 3.125 mL. **Answer:** 3.125 mL

9. 0.1% w/v solution contains 100 mg per 100 mL. 500 mL of a 0.1% w/v solution contains 100 × 5 = 500 mg. 20% w/v solution contains 20 g per 100 mL. Therefore a 20% w/v solution will contain 2 g per 10 mL, 1 g per 5 mL and 500 mg per 2.5 mL. **Answer:** 2.5 mL of a 20% solution would be required to make up 500 mL of a 0.1% w/v solution.

10. 0.9% w/v solution contains 900 mg per 100 mL. Therefore, we require (900 ÷ 100) × 5000 = 45 000 mg = 45 g. **Answer:** 45 g

11. 5% = 5 g per 100 mL. 100 mL ÷ 20 mL = 5.5 ÷ 5 = 1 g. **Answer:** 1 g

12. 0.2% w/v solution contains 200 mg per 100 mL. Therefore, there are 100 mg per 50 mL. **Answer:** d (100 mg)

13. 5% w/v solution contains 5 g per 100 mL. Therefore, there is 1 g per 20 mL. **Answer:** c (1000 mg (or 1 g))

14. 0.01% w/v solution contains 0.01 g = 10 mg per 100 mL. 10 × 3 = 30 mg per 300 mL. **Answer:** c (30 mg)

15. 15% w/v solution contains 15 g per 100 mL. (15 ÷ 100) × 750 = 112.5 g **Answer:** d (112.5 g)

16. 0.45% w/v solution contains 450 mg per 100 mL. Therefore, we require (450 ÷ 100) × 10 000 = 45 000 mg = 45 g. **Answer:** d (45 g)

17. 5 mL per 1000 mL is the same as 0.5 mL per 100 mL. 0.5 mL per 100 mL = 0.5% v/v. **Answer:** c (0.5% v/v)

18. 5% is equivalent to 5 g per 100 mL. As the prescription requires 10 mL, we require 5 ÷ 10 = 0.5 g. **Answer:** c (0.5 g)

19. 0.5% w/v solution contains 500 mg (0.5 g) per 100 mL. 250 mL will contain (500 ÷ 100) × 250 = 1250 mg (1.25 g). 20% w/v solution contains 20 g per 100 mL, 2 g per 10 mL or 1 g per 5 mL. Therefore, we require (1.25 ÷ 1) × 5 = 6.25 mL. **Answer:** b (6.25 mL)

20. 1% w/v solution contains 1 g per 100 mL. Therefore, we require (1 ÷ 100) × 150 = 1.5 g. 4% w/v solution contains 4 g per 100 mL. This is equal to 2 g per 50 mL, 1 g per 25 mL and 0.5 g per 12.5 mL. Therefore, 1.5 g will be contained in (1.5 ÷ 0.5) × 12.5 = 37.5 mL. **Answer:** e (37.5 mL)

21. 0.5% w/v contains 500 mg per 100 mL. 250 mL of a 0.5% w/v solution contains (500 ÷ 100) × 250 = 1250 mg (1.25 g). 25% w/v solution contains 25 g per 100 mL. Therefore 1.25 g will be contained in (1.25 ÷ 25) × 100 = 5 mL. **Answer:** d (5 mL)

22.

a. A 1% v/v solution contains 1 mL per 100 mL. Therefore half a litre (500 mL) will contain 1 × 5 = 5 mL. A 15% v/v solution will contain 15 mL per 100 mL. Therefore for 5 mL, we require 100 ÷ (15 ÷ 5) = 33.3 mL. **Answer:** 33.3 mL

b. A 1% v/v solution contains 1 mL per 100 mL. Therefore 250 mL will contain (1 ÷ 100) x 250 = 2.5 mL. A 40% v/v solution will contain 40 mL per 100 mL. Therefore for 2.5 mL, we require 100 ÷ (40 ÷ 2.5) = 6.25 mL. **Answer:** 6.25 mL.

c. A 1% v/v solution contains 1 mL per 100 mL. 500 mL of a 1% v/v solution contains 5 mL. A 10% solution contains 10 mL per 100 mL. Therefore for 5 mL, we require 100 ÷ (10 ÷ 5) = 50 mL. **Answer:** 50 mL.

d. A 0.5% v/v solution contains 0.5 mL per 100 mL. 1 litre (1000 mL) of a 0.5% v/v solution contains 5 mL. A 15% solution contains 15 mL per 100 mL. Therefore for 5 mL, we require 100 ÷ (15 ÷ 5) = 33.3 mL. **Answer:** 33.3 mL.

e. A 0.05% v/v solution contains 0.05 mL per 100 mL. 1 litre (1000 mL) of a 0.05% v/v solution contains 0.5 mL (0.05 × 10). A 4% solution contains 4 mL per 100 mL. Therefore for 0.5 mL, we require 100 ÷ (4 ÷ 0.5) = 12.5 mL. **Answer:** 12.5 mL.

23. If a solid is soluble in 2.5 parts of water, that means that 1 g is soluble in 2.5 mL of water. Therefore, for 3 g, we require (3 ÷ 1) × 2.5 = 7.5 mL. **Answer:** d (7.5 mL)

24. If a solid is soluble 1 in 1.5 parts of water, that means that 1 g is soluble in 7.5 mL of water. Therefore, for 7 g, we require (7 ÷ 1) × 1.5 = 10.5 mL. **Answer:** d (10.5 mL)

25. If sodium bicarbonate is soluble 1 in 11, that means that 1 g is soluble in 11 mL. Therefore, for 0.37 kg (or 370 g), we require 370 × 11 = 4070 mL or 4.07 litres. **Answer:** b (4.07 L)

26. A 0.2% solution contains 200 mg per 100 mL. 200 mL of a 0.2% solution contains 400 mg. A 1 in 150 w/v solution contains 1 g in 150 mL. Therefore for 400 mg, we require (150 ÷ 1) × 0.4 = 60 mL. **Answer:** e (60 mL)

27. 500 micrograms in 2 mL is equivalent to 1 mg per 4 mL. This is equivalent to 1 g (1 mg × 1000) in 4000 mL (4 × 1000). **Answer:** e (1 in 4000)

28. A 1 in 12 000 solution contains 1 g in 12 000 mL (12 litres). Therefore, for 5.4 litres, we will require (1 ÷ 12) × 5.4 = 0.45 g (450 mg). **Answer:** d (450 mg)

29. 1 litre of a molar solution contains 1 mol of sodium chloride per litre. Therefore, there are 58.44 g of sodium chloride per litre. **Answer**: e (58.44 g)

30. 1 mol of sodium bicarbonate contains 84 g. Therefore, 1 mmol contains 0.084 g (84 mg) sodium bicarbonate. Therefore, 0.5 mmol contains 42 mg. Therefore 150 mL of a 0.5 mmol/mL solution contains 150 × 42 = 6300 mg (6.3 g). **Answer**: c (6.3 g)

31. 1 mol of Sodium Chloride BP contains 58.44 g. Therefore, 1 mmol contains 0.05844 g (58.44 mg). Therefore, 2 mmol contains 116.88 mg. Therefore, 100 mL of a 2 mmol/mL solution contains 116.88 × 100 = 11 688 mg (11.688 g). **Answer**: c (11.69 g)

32. 1 mol of Sodium Bicarbonate BP contains 84 g. Therefore, 1 mmol contains 0.084 g (84 mg) Sodium Bicarbonate BP. Therefore 75 mL of a 1 mmol/mL solution contains 75 × 84 = 6300 mg (6.3 g). **Answer**: c (6.3 g)

33. Sodium Bicarbonate Solution 0.5 mmol/mL

   1. Used to treat chronic acidotic states (*British National Formulary* 61st edn, p 596).

   2. This patient has been recently discharged from hospital and the hospital states that the recommended dose for a child 1 month–2 years is 1–2 mmol/kg daily in 1–2 divided doses. In order to confirm the dose on the prescription is safe and suitable, the weight of the child would be required. The parent would be able to provide this information: this would be more accurate than the average weight listed in the *British National Formulary*. In this case the child weighs 5 kg, therefore the dose that would be safe is 5–10 mmol daily in divided doses and the amount ordered on the prescription concurs with this.

   3. In order to prepare the product formula, we need to consider the molecular weight of Sodium Bicarbonate BP, which is 84.

       1 mol Sodium Bicarbonate BP = 84 g
       1 mol = 1000 mmol = 84 g
       200 mL of a 0.5 mmol solution is required
       Therefore 200 x 0.5 mmol required = 100 mmol
       100 mmol = 84/1000 x 100 = 8.4 g Sodium Bicarbonate BP

	Master	200 mL
Sodium Bicarbonate BP	4.2 g	8.4 g
Freshly boiled and cooled purified water	to 100 mL	to 200 mL

   4.

   a. Sodium Bicarbonate BP is soluble 1 in 11 in water. Therefore to dissolve 8.4 g Sodium Bicarbonate BP a minimum of 8.4 x 11 = 92.4 mL of water would be required.

   b. Freshly boiled and cooled purified water would be used as the solution is intended to be administered to a baby with renal problems.

   c. None added.

   d. Not appropriate.

The following method would be used to prepare 200 mL of sodium bicarbonate solution 0.5 mmol/mL from the formula above:

1. Weigh 8.4 g Sodium Bicarbonate BP on a Class II or electronic balance.
2. Measure approximately 100 mL of freshly boiled and cooled purified water and transfer to a beaker.
3. Add the Sodium Bicarbonate BP to the water in the beaker.
4. Stir to aid dissolution.
5. Transfer to a 250 mL conical measure with rinsings.
6. Make up to volume with freshly boiled and cooled purified water.
7. Transfer to a 200 mL amber flat medical bottle with a child-resistant closure and label.

5. A plain amber bottle with a child-resistant closure would be most suitable as the preparation is a solution for internal use.

6.

a. The product is unofficial, therefore the following title would be suitable: 'Sodium Bicarbonate BP solution 0.5 mmol/mL'.

b. The product is unofficial, therefore it is necessary to put the quantitative particulars on the label. As the product is intended for internal use, the quantitative particulars would be expressed per dose.

Each 10 mL dose contains:

Sodium Bicarbonate BP	420 mg
Freshly boiled and cooled purified water	to 10 mL

or

Each 10 mL dose contains:

Sodium Bicarbonate BP	5 mmol
Freshly boiled and cooled purified water	to 10 mL

c. Not applicable.
d. 'Give 10 mL using the oral syringe provided TWICE daily with feeds.'
e. Not applicable.
f. 2 weeks as there is no preservative present

g.

Sodium Bicarbonate BP Solution 0.5mmol/mL	**200 mL**
Give 10 mL using an oral syringe TWICE daily with feeds.	
Do not use after (2 weeks)	
Each 10 mL contains:	
Sodium Bicarbonate BP	420 mg
Freshly boiled and cooled purified water	to 10 mL
Miss Julie Jordan	Date of dispensing

7. The parent/guardian would be shown how to use the oral syringe and instructed to give 10 mL using the oral syringe twice daily with feeds. In addition, the discard date would be highlighted to the parent/guardian

34. Hibitane solution 0.05%

1. Used for general skin disinfection (*British National Formulary* 61st edn, p 743).
2. The recommended strength for this purpose is 0.05% or 1 in 100 dilution of 5% concentrate (*British National Formulary* 61st edn, p 743).
3. In order to prepare this product we need to dilute a concentrate that is commercially available. Hibitane Concentrate 5% is the strength easily obtainable.

   0.05% means 50 mg per 100 mL, therefore 75 mg per 150 mL
   Our concentrate contains 5 g in 100 mL
   Therefore 500 mg in 10 mL. Therefore 50 mg in 1 mL
   Therefore we need 1.5 mL = 75 mg

	Master	50 mL	150 mL
Hibitane Concentrate 5%	1 mL	0.5 mL	1.5 mL
Freshly boiled and cooled purified water	to 100 mL	to 50 mL	to 150 mL

4.

a. The Hibitane concentrate is miscible with both water and alcohol (*British National Formulary* 51st edn, p 607).
b. Freshly boiled and cooled purified water would be the vehicle of choice as the preparation is being used as a skin disinfectant.
c. Not applicable.
d. Not applicable.

   The following method would be used to prepare 150 mL of Hibitane solution 0.05% from the formula above:

   1. Measure 1.5 mL Hibitane Concentrate 5% using a syringe.
   2. Measure approximately 100 mL of freshly boiled and cooled purified water and transfer to a beaker.
   3. Add the Hibitane Concentrate 5% to the water in the beaker.
   4. Stir to aid dissolution.
   5. Transfer to a 250 mL conical measure with rinsings.
   6. Make up to volume with freshly boiled and cooled purified water.
   7. Transfer to a 150 mL fluted amber medical bottle with a child-resistant closure and label.

5. A fluted amber bottle with a child-resistant closure would be most suitable as the preparation is a solution for external use.

6.

a. The product is unofficial, therefore the following title would be suitable: 'Hibitane Solution 0.05%'.
b. The product is unofficial, therefore it is necessary to put the quantitative particulars on the label. As the product is intended for external use, the quantitative particulars would be expressed per container.

   Each 150 mL contains:

Hibitane Concentrate 5%	1.5 mL
Freshly boiled and cooled purified water	to 150 mL

    c. 'For external use only' will need to be added to the label as the product is a solution for external use.

    d. 'Use for cleansing the affected area as directed by your doctor.'

    e. Not applicable.

    f. Two weeks as there is no preservative and the product may be applied to open wounds.

    g.

Hibitane Solution 0.05%	**150 mL**

Use for cleansing the affected area as directed by your doctor.

For external use only

Do not use after (2 weeks)

Each 150 mL contains:

Hibitane Concentrate 5%	1.5 mL
Freshly boiled and cooled purified water	to 150 mL
Mrs Sally Burns	Date of dispensing

7. The patient would be advised to use the solution for cleansing the affected area as directed by her doctor. In addition, the discard date and the fact that the product is for external use only would be highlighted to the patient.

Chapter 3: Suspensions

1. Diffusible solids are powders that are light and wettable. They mix readily with water and on shaking diffuse evenly through the liquid and remain in suspension long enough for a dose to be accurately measured. Indiffusible solids are insoluble powders which will not remain evenly distributed in a vehicle long enough to ensure uniformity of dose. This may be corrected by increasing the viscosity by adding a thickening/suspending agent.

2. Chalk BP is an indiffusible powder (see Example 3.2). **Answer:** a (Chalk BP)

3. Magnesium Trisilicate BP is a diffusible powder (see Example 3.1). **Answer:** d (Magnesium Trisilicate BP)

4. Bentonite BP would be unsuitable for use for a suspension intended for the oral route of administration. **Answer:** c (Bentonite BP)

5. **Answer:** b (Zinc Sulphate Lotion BP)

6. The formula for Compound Tragacanth Powder BP is:

Acacia BP	20%
Tragacanth BP	15%
Starch BP	20%
Sucrose BP	45%

7. Compound Tragacanth Powder BP is unsuitable for external use as the sucrose content will render any suspension sticky and therefore unsuitable for external application and pharmaceutically inelegant.

8. 'Shake the bottle.'

9. Quantitative particulars are expressed per dose for internal products. Therefore, in this case, quantitative particulars would be expressed per 15 mL. **Answer**: c (Amount per 15 mL)

10. The patient needs to take 1 g of paracetamol per dose. The stock suspension contains 120 mg per 5 mL. Therefore, each dose requires (1000 ÷ 120) × 5 = 41.67 mL. For ease of administration, the patient would be advised to take 41.7 mL per dose.

    The patient is required to take each dose four times a day and a 2-week supply has been requested. Therefore, you need to dispense 41.7 × 4 × 14 = 2335.2 mL. **Answer**: e (2.4 litres)

11. 200 mg of Tragacanth BP is used per 100 mL of indiffusible suspension. Therefore, 400 mg of Tragacanth BP would be required (200 mg × 2). **Answer**: c (400 mg)

12. Calamine BP would be included at a concentration of 15%, i.e. 15 g per 100 mL of suspension.

    Zinc Oxide BP would be included at a concentration of 5%, i.e. 5 g per 100 mL of suspension.

    Bentonite BP is chosen as the suspending agent as the suspension is for external use (a lotion). Bentonite BP would be used in a concentration of 2–3%.

    Freshly boiled and cooled purified water would be chosen as the vehicle as the suspension does not contain a preservative.

    Therefore, the formula would be:

Calamine BP	15 g
Zinc Oxide BP	5 g
Bentonite BP	2 g or 3 g
Freshly boiled and cooled purified water	to 100 mL

13. Compound Tragacanth Powder BP is used as 2 g per 100 mL of final suspension. Therefore you will require (150 ÷ 100) × 2 = 3 g. **Answer**: d (3 g)

14. **Answer**: c (All suspensions need a direction to shake the bottle.)

15. **Answer**: b (The amount of suspending agent used depends on the amount of powder to be suspended.)

16. Magnesium Carbonate Mixture BPC

    1. Used as an antacid (*British National Formulary* 61st edn, p 45).
    2. This is an official preparation, therefore the formula is safe and suitable for purpose. The dose of 10 mL three times a day is consistent with the recommended dosage (*British Pharmaceutical Codex* 1973, p 746).
    3. Prepare 150 mL of Magnesium Carbonate Mixture BPC.

Product formula
(from the *British Pharmaceutical Codex* 1973, p 746)

	Master	100 mL	50 mL	150 mL
Light Magnesium Carbonate BP	50 g	5 g	2.5 g	7.5 g
Sodium Bicarbonate BP	80 g	8 g	4 g	12 g
Concentrated Peppermint Emulsion BP	25 mL	2.5 mL	1.25 mL	3.75 mL
Double Strength Chloroform Water BP	500 mL	50 mL	25 mL	75 mL
Water	to 1000 mL	to 100 mL	to 50 mL	to 150 mL

Interim formula for Double Strength Chloroform Water BP

Concentrated Chloroform Water BPC 1959	5 mL
Potable water	to 100 mL

4.

a. Sodium Bicarbonate BP is soluble 1 in 11 in water (*British Pharmacopoeia* 1988, p 509). Therefore to dissolve 12 g Sodium Bicarbonate BP a minimum of 12 x 11 = 132 mL of water would be required.

b. Double Strength Chloroform Water BP and potable water would be used as the vehicle as per the product formula.

c. Double Strength Chloroform Water BP is included in this product as the preservative as per the product formula.

d. Concentrated Peppermint Emulsion BP is added as a flavouring as per the product formula.

The following method would be used to prepare 150 mL of Magnesium Carbonate Mixture BPC from the formula above:

1. Using the master formula from the *British Pharmaceutical Codex* for 1000 mL of final product, calculate the quantity of ingredients required to produce the final volume needed (150 mL).

2. Calculate the composition of a convenient quantity of Double Strength Chloroform Water BP, sufficient to satisfy the formula requirements but also enabling simple, accurate measurement of the concentrated component.

Method of compounding for Double Strength Chloroform Water BP

a. In this case, 75 mL of Double Strength Chloroform Water BP is required and so it would be sensible to prepare 100 mL. To prepare 100 mL Double Strength Chloroform Water BP, measure 5 mL of Concentrated Chloroform Water BPC 1959 accurately using a 5 mL conical measure.

b. Add approximately 90 mL of potable water to a 100 mL conical measure (i.e. sufficient water to enable dissolution of the concentrated chloroform component without reaching the final volume of the product).

c. Add the measured Concentrated Chloroform Water BPC 1959 to the water in the conical measure.

d. Stir gently and then accurately make up to volume with potable water.

e. Visually check that no undissolved chloroform remains at the bottom of the measure.

Noting that sodium bicarbonate is soluble 1 in 11 with water, a minimum of 11 mL of water would be required to dissolve 1 g of sodium bicarbonate. The final volume of Magnesium Carbonate Mixture BPC required (150 mL) will contain 12 g of Sodium Bicarbonate BP. As 1 g of sodium bicarbonate is soluble in 11 mL, 12 g is soluble in 132 mL (12 x 11 = 132 mL). Therefore a minimum of 132 mL of vehicle would be required to dissolve the 12 g of sodium bicarbonate in this example. For ease of compounding choose a convenient volume of vehicle, say 135 mL, in which to dissolve the solute initially. When choosing the amount of vehicle to use for dissolution, it is important to consider the total amount of each liquid ingredient in the preparation to ensure that only the correct amounts are added or the final product does not go over volume.

3. Weigh 7.5 g of Light Magnesium Carbonate BP on a Class II or electronic balance.
4. Weigh 12 g Sodium Bicarbonate BP on a Class II or electronic balance.
5. Measure 3.75 mL of Concentrated Peppermint Emulsion BP using a 1 mL syringe and a 5 mL syringe.
6. Measure 75 mL Double Strength Chloroform Water BP and transfer to a beaker.
7. Add approximately 60 mL of potable water to the Double Strength Chloroform Water BP in the beaker in order to produce 135 mL of vehicle (in order to produce sufficient volume to dissolve the 12 g Sodium Bicarbonate BP).
8. Transfer the Sodium Bicarbonate BP to the beaker and stir to aid dissolution.
9. Transfer the Light Magnesium Carbonate BP to a porcelain mortar.
10. Add a small amount of the sodium bicarbonate solution to the powder in the mortar to make a smooth paste.
11. Slowly add the solution of sodium bicarbonate, mixing with a pestle to form a smooth pourable paste.
12. Transfer the contents of the mortar to a 250 mL conical measure.
13. Rinse out the mortar with more solution and add the rinsings to the conical measure.
14. Add the Concentrated Peppermint Emulsion BP to the mixture in the conical measure.
15. Make up to volume with any remaining solution and potable water.
16. Stir and transfer to an amber flat medical bottle, label and dispense.

5. A plain amber bottle with a child-resistant closure would be most suitable as the preparation is a solution for internal use.
6.
a. The product is official, therefore the following title would be suitable: 'Magnesium Carbonate Mixture BPC'.
b. Quantitative particulars are not required as the product is official.
c. 'Shake the bottle' will need to be added to the label as the product is a suspension and will need shaking before use to ensure an accurate dose is measured.
d. 'Take TWO 5 mL spoonfuls THREE times a day in water.'
e. Not applicable.
f. This product would be recently prepared as Double Strength Chloroform Water BP acts as a preservative and therefore would attract a 4-week discard date.

g.

Magnesium Carbonate Mixture BPC	**150 mL**
Take TWO 5 mL spoonfuls THREE times a day in water.	
Do not use after (4 weeks)	
Shake the bottle	
Mrs Sally Marlow	Date of dispensing

7.   The patient would be advised to mix two 5 mL spoonfuls with an equal volume of water and take three times a day. In addition, the discard date and the need to shake the bottle would be highlighted.

17. Zinc and talc lotion
1.   Mildly astringent, used topically as a soothing and protective application (*Martindale* 35th edn, pp 1453, 1458 and 2100).
2.   This is an unofficial preparation, therefore the formula will need to be checked to ensure that it is safe and suitable for purpose. The formula is comparable to official products containing the same ingredients listed in the same or similar proportions (*Martindale* 35th edn, pp 1453, 1458 and 2100). Therefore the product is safe and suitable for the intended purpose.
3.   Bentonite BP is used for external suspensions in a concentration of 2–3%. Therefore, for 100 mL of suspension, it would be appropriate to use 2 g of Bentonite BP.

Prepare 50 mL of zinc and talc lotion.

Product formula

	Master	**100 mL**	**50 mL**
Purified Talc BP	25%	25 g	12.5 g
Zinc Oxide BP	25%	25 g	12.5 g
Glycerin BP	5%	5 mL	2.5 mL
Bentonite BP	qs	2 g	1 g
Water	to 100%	to 100 mL	to 50 mL

4.
a.   Purified Talc BP (*British Pharmacopoeia* 1988, p 552), Zinc Oxide BP (*British Pharmacopoeia* 1988, p 604) and Bentonite BP (*British Pharmacopoeia* 1988, p 62) are all practically insoluble in water.
b.   Freshly boiled and cooled purified water would be the vehicle of choice as the product does not contain a preservative.
c.   No preservative is included in this product.
d.   The product is for external use and therefore does not require the addition of a flavouring.

The following method would be used to prepare 50 mL of zinc and talc lotion from the formula above:

1. Weigh 1 g Bentonite BP on a Class II or electronic balance.
2. Transfer to a porcelain mortar.
3. Weigh 12.5 g Zinc Oxide BP on a Class II or electronic balance.
4. Add the Zinc Oxide BP to the Bentonite BP in the mortar using the doubling-up method.
5. Weigh 12.5 g Purified Talc BP on a Class II or electronic balance and add to the mortar using the doubling-up method.
6. Measure 2.5 mL Glycerin BP using a suitable syringe
7. Add the Glycerin BP to the powders in the mortar and mix to form a paste.
8. Add some freshly boiled and cooled purified water to the powders to make a pourable paste.
9. Transfer the pourable paste to a 50 mL conical measure.
10. Rinse out the mortar using freshly boiled and cooled purified water and add the rinsings to the final measure.
11. Make up to volume with freshly boiled and cooled purified water and stir.
12. Transfer to an amber fluted bottle and label.

5. An amber fluted bottle with a child-resistant closure would be most suitable as the preparation is for external use.

6.

a. The product is unofficial therefore the following title would be suitable: 'The Lotion'.

b. Quantitative particulars are required as the product is unofficial. The quantitative particulars would be expressed per container:

Each 50 mL contains:

Zinc Oxide BP	25%
Purified Talc BP	25%
Glycerin BP	5%
Bentonite BP	2%
Freshly boiled and cooled purified water	to 100%

c. 'For External Use Only' will need to be added to the label as the product is for external use.
'Shake the bottle' will need to be added to the label as the product is a suspension and will need shaking before use to ensure an accurate dose is measured.

d. 'Apply when required for itching.'

e. Not applicable.

f. This product would attract a 2-week discard date as it does not contain a preservative.

g.

The Lotion	50 mL

Apply when required for itching.
Discard after (2 weeks)
For external use only
Shake the bottle

Each 50 mL contains:

Zinc Oxide BP	25%
Purified Talc BP	25%
Glycerin BP	5%
Bentonite BP	2%
Freshly boiled and cooled purified water	to 100%

Mr Edward Smith   Date of dispensing

7. The patient would be advised to apply the lotion when required for itching. In addition, the discard date, the need to shake the bottle before use and the fact that the product is for external use only would be highlighted to the patient.

Chapter 4: Emulsions

1. An emulsion is essentially a liquid preparation containing a mixture of oil and water that is rendered homogeneous by the addition of an emulsifying agent.
2. It is important to include a preservative in an oral emulsion to prevent microbial growth as emulsions are particularly susceptible to microbial growth.
3. As emulsions are particularly susceptible to microbial contamination, freshly boiled and cooled purified water is used to minimise the risk of contamination.
4. The complete table is:

	Proportion in primary emulsion		
Type of oil	Oil	Aqueous	Gum
Fixed oil	4	2	1
Volatile oil	2	2	1
Mineral oil	3	2	1

5. **Answer:** d (Liquid Paraffin BP) (see Example 4.3)
6. **Answer:** b (Chlorocresol BP)
7. **Answer:** c (Peppermint Oil BP)
8. **Answer:** a (4:2:1)
9. **Answer:** a (A creamed emulsion will reform on shaking.)
10. Liquid Paraffin BP is a mineral oil. Therefore, the proportions of oil to aqueous to gum is 3:2:1. 36% v/v of 100 mL is 36 mL. Therefore, within the primary emulsion, 3 parts is equal to 36. Therefore 1 part of Acacia BP is equal to 12 g (36 ÷ 3). **Answer:** c (12 g)
11. If the final volume of emulsion is 100 mL, the master formula of the product will contain 50 mL of Double Strength Chloroform Water BP (i.e. 50% of the final volume). (Please note that 'b' (24 mL) is the amount of aqueous phase (i.e. Double Strength Chloroform Water BP) that will be used in the primary emulsion.) **Answer:** d (50 mL)

12. Peppermint Oil BP is a volatile oil. Therefore, the formula for the primary emulsion is 2:2:1. The final product will contain 20% v/v Peppermint Oil BP, i.e. 20 mL of Peppermint Oil BP per 100 mL of final product. If the volume of product required is 50 mL, you will require 10 mL of Peppermint Oil BP. Therefore, within the primary emulsion, 2 parts is equal to 10. Therefore 1 part of Acacia BP is equal to 5 g (10 ÷ 2). **Answer:** b (5 g)

13. **Answer:** c (4 weeks)

14. **Answer:** c (Shake the bottle)

15. Maize oil 30% v/v emulsion
    1. Used as a high-calorie nutritional supplement (*Martindale* 35th edn, p 1793).
    2. The product is being used as a food supplement and so is safe and suitable for use.
    3. Prepare 100 mL of maize oil emulsion 30% v/v.

Product formula

Maize Oil BP	**30 mL**
Acacia BP	qs
Double Strength Chloroform Water BP	50 mL
Freshly boiled and cooled purified water	to 100 mL

The quantity of emulsifying agent (Acacia BP) required to produce 100 mL of the emulsion must be calculated.

Formula for primary emulsion

Maize oil is a fixed oil. Therefore the primary emulsion ratio is:

Oil : Water : Gum

4 : 2 : 1

30 mL of Maize Oil BP is required, therefore 4 parts = 30 mL. 1 part will therefore be 30 ÷ 4 = 7.5.

Therefore:

The amount of freshly boiled and cooled purified water needed = 2 × 7.5 mL = 15 mL.

The amount of acacia required = 7.5 g.

Therefore the product formula for 100 mL of maize oil 30% emulsion is:

	100 mL
Maize Oil BP	30 mL
Acacia BP	7.5 g
Double Strength Chloroform Water BP	50 mL
Freshly boiled and cooled purified water	to 100 mL

Interim formula for Double Strength Chloroform Water BP

Concentrated Chloroform Water BPC 1959	2.5 mL
Freshly boiled and cooled purified water	to 50 mL

4.

a. No solids will need to be dissolved during the preparation of this product.

b. As emulsions are particularly susceptible to microbial contamination, Double Strength Chloroform Water BP will be used as the vehicle at a concentration of 50%. Freshly boiled and cooled purified water will be used as the remainder of the vehicle. As freshly boiled and cooled purified water is used in the product, it will also be used to make the Double Strength Chloroform Water BP.

c. Double Strength Chloroform Water BP is included in this product as the preservative as per the product formula.

d. No extra flavouring is required. In addition to preservative action, Double Strength Chloroform Water BP will give some flavouring.

The following method would be used to prepare 100 mL of maize oil 30% v/v emulsion from the formula above:

1. Calculate the composition of a convenient quantity of Double Strength Chloroform Water BP, sufficient to satisfy the formula requirements but also enabling simple, accurate measurement of the concentrated component.

Method of compounding for Double Strength Chloroform Water BP

a. In this case, 50 mL of Double Strength Chloroform Water BP is required. To prepare 50 mL Double Strength Chloroform Water BP, measure 2.5 mL of Concentrated Chloroform water BPC 1959 accurately using a 5 mL and a 1 mL syringe.

b. Add approximately 45 mL of freshly boiled and cooled purified water to a 50 mL conical measure (i.e. sufficient water to enable dissolution of the concentrated chloroform component without reaching the final volume of the product).

c. Add the measured Concentrated Chloroform Water BPC 1959 to the water in the conical measure.

d. Stir gently and then accurately make up to volume with freshly boiled and cooled purified water.

e. Visually check that no undissolved chloroform remains at the bottom of the measure.

2. Measure 30 mL of Maize Oil BP.

3. Transfer to a clean dry porcelain mortar.

4. Weigh 7.5 g Acacia BP on a Class II balance.

5. Transfer to the mortar and mix gently (approx 3 stirs) to wet the acacia in the oil.

6. Measure 50 mL Double Strength Chloroform Water BP.

7. Measure 15 mL of Double Strength Chloroform Water BP (from the 50 mL in step 6) in an appropriate measure and add to the mortar in one go.

8. Stir vigorously using the pestle in *one* direction until the primary emulsion is formed.

9. Add more Double Strength Chloroform Water BP little by little until the emulsion is pourable.

10. Transfer to an appropriate conical measure with rinsings.

11. Make up to volume with any remaining Double Strength Chloroform Water BP and freshly boiled and cooled purified water.

12. Stir and transfer to an amber flat medical bottle, label and dispense.

5. A plain amber bottle with a child-resistant closure would be most suitable as the preparation is for internal use.

6.

a. The product is unofficial, therefore the following title would be suitable: 'Maize oil 30% v/v emulsion'.

b. Quantitative particulars are required as the product is unofficial. The quantitative particulars would be expressed per dose (i.e. per 15 mL):

Each 15 mL contains:
Maize oil BP	4.5 mL
Acacia BP	1.125 g
Double Strength Chloroform Water BP	7.5 mL
Freshly boiled and cooled purified water	to 15 mL

c. 'Shake the bottle' will need to be added to the label as the product is a emulsion and will need shaking before use to ensure an accurate dose is measured.

d. 'Take THREE 5 mL spoonfuls THREE times a day.'

e. Not applicable.

f. This product is an emulsion and would attract a 4-week discard date.

g.

Maize oil 30% v/v emulsion	**100 mL**
Take THREE 5 mL spoonfuls THREE times a day.	
Do not take after (4 weeks)	
Shake the bottle	
Each 15 mL contains:	
Maize Oil BP	4.5 mL
Acacia BP	1.125 g
Double Strength Chloroform Water BP	7.5 mL
Freshly boiled and cooled purified water	to 15 mL
Miss Sophie Jones	Date of dispensing

7. The patient would be advised to take three 5 mL spoonfuls three times a day. In addition, the discard date and the need to shake the bottle before use would be highlighted to the patient.

16. Castor oil 30% v/v emulsion

1. It was previously used as a powerful stimulant laxative. Although now considered obsolete (*British National Formulary* 61st edn, p 68), it may be used in exceptional circumstances.

2. This is an unofficial preparation, therefore the formula will need to be checked to ensure that it is safe and suitable for purpose. The dose for Castor Oil BP is 5–20 mL (*Martindale* 26th edn, p 1256). The dose requested on this prescription falls within this range and so the product is safe and suitable for use.

3. Prepare 50 mL of castor oil emulsion 30% v/v.

Product formula

	Master	50 mL
Castor Oil BP	30 mL	15 mL
Acacia BP	qs	qs
Double Strength Chloroform Water BP	50 mL	25 mL
Freshly boiled and cooled purified water	to 100 mL	to 50 mL

Need to calculate the quantity of emulsifying agent (Acacia BP) required to produce 50 mL of the emulsion.

Formula for primary emulsion
Castor oil is a fixed oil. Therefore the primary emulsion ratio is:

Oil : Water : Gum
4 : 2   : 1

15 mL of Castor Oil BP is required, therefore 4 parts = 15 mL. 1 part will therefore be 15 ÷ 4 = 3.75.

Therefore:
The amount of freshly boiled and cooled purified water needed = 2 x 3.75 mL = 7.5 mL.
The amount of acacia required = 3.75 g.
Therefore the product formula for 50 mL of castor oil 30% emulsion is:

	50 mL
Castor Oil BP	15 mL
Acacia BP	3.75 g
Double Strength Chloroform Water BP	25 mL
Freshly boiled and cooled purified water	to 50 mL

Interim formula for Double Strength Chloroform Water BP

Concentrated Chloroform Water BPC 1959	2.5 mL
Freshly boiled and cooled purified water	to 50 mL

4.
a. No solids will need to be dissolved during the preparation of this product.
b. As emulsions are particularly susceptible to microbial contamination, Double Strength Chloroform Water BP will be used as the vehicle at a concentration of 50%. Freshly boiled and cooled purified water will be used as the remainder of the vehicle. As freshly boiled and cooled purified water is used in the product, it will also be used to make the Double Strength Chloroform Water BP.
c. Double Strength Chloroform Water BP is included in this product as the preservative as per the product formula.
d. No extra flavouring is required. In addition to preservative action Double Strength Chloroform Water BP will give some flavouring.

The following method would be used to prepare 50 mL of castor oil 30% v/v emulsion from the formula above:
1. Calculate the composition of a convenient quantity of Double Strength Chloroform Water BP, sufficient to satisfy the formula requirements but also enabling simple, accurate measurement of the concentrated component.

Method of compounding for Double Strength Chloroform Water BP

a. In this case, 25 mL of Double Strength Chloroform Water BP is required and so it would be sensible to prepare 50 mL. To prepare 50 mL Double Strength Chloroform Water BP, measure 2.5 mL of Concentrated Chloroform Water BPC 1959 accurately using a 5 mL and a 1 mL syringe.
b. Add approximately 45 mL of freshly boiled and cooled purified water to a 50 mL conical measure (i.e. sufficient water to enable dissolution of the concentrated chloroform component without reaching the final volume of the product).
c. Add the measured Concentrated Chloroform Water BPC 1959 to the water in the conical measure.
d. Stir gently and then accurately make up to volume with freshly boiled and cooled purified water.
e. Visually check that no undissolved chloroform remains at the bottom of the measure.

2. Measure 15 mL of Castor Oil BP.
3. Transfer to a clean dry porcelain mortar.
4. Weigh 3.75 g Acacia BP on a Class II balance.
5. Transfer to the mortar and mix gently (approx 3 stirs) to wet the acacia in the oil.
6. Measure 25 mL Double Strength Chloroform Water BP.
7. Measure 7.5 mL of Double Strength Chloroform Water BP (from the 25 mL in step 6) in an appropriate measure and add to the mortar in one go.
8. Stir vigorously using the pestle in *one* direction until the primary emulsion is formed.
9. Add more Double Strength Chloroform Water BP little by little until the emulsion is pourable.
10. Transfer to an appropriate conical measure with rinsings.
11. Make up to volume with any remaining Double Strength Chloroform Water BP and freshly boiled and cooled purified water.
12. Stir and transfer to an amber flat medical bottle, label and dispense.

5. A plain amber bottle with a child-resistant closure would be most suitable as the preparation is for internal use.

6.

a. The product is unofficial, therefore the following title would be suitable: 'Castor oil 30% v/v emulsion'.
b. Quantitative particulars are required as the product is unofficial. The quantitative particulars would be expressed per dose (i.e. per 20 mL):

Each 20 mL contains:

Castor Oil BP	6 mL
Acacia BP	1.5 g
Double Strength Chloroform Water BP	10 mL
Freshly boiled and cooled purified water	to 20 mL

c.   'Shake the bottle' will need to be added to the label as the product is a emulsion and will need shaking before use to ensure an accurate dose is measured.

d.   'Take FOUR 5 mL spoonfuls at night.'

e.   Not applicable.

f.   This product is an emulsion and would attract a 4-week discard date.

g.

Castor oil 30% v/v emulsion	50 mL

Take FOUR 5 mL spoonfuls at night.

Do not take after (4 weeks)

Shake the bottle

Each 20 mL contains:

Castor Oil BP	6 mL
Acacia BP	15 g
Double Strength Chloroform Water BP	10 mL
Freshly boiled and cooled purified water	to 20 mL
Ms Petra Williams	Date of dispensing

7.   The patient would be advised to take four 5 mL spoonfuls at night. In addition, the discard date and the need to shake the bottle before use would be highlighted to the patient.

Chapter 5: Creams

1.   **Answer:** c (The surfactant in creams allows better skin penetration than in ointments.)

2.   A 1 in 20 dilution will contain 1 g of Betnovate Cream per 20 g of dilution. Therefore, in 200 g of dilution, there will be (200 ÷ 20) × 1 = 10 g. **Answer:** b (10 g)

3.   A 1 in 4 dilution will contain 1 g of Eumovate Cream per 4 g of dilution. Therefore, in 200 g of dilution, there will be (200 ÷ 4) × 1 = 50 g. **Answer:** d (50 g)

4.   A 1 in 5 dilution will contain 1 g of Betnovate Cream per 5 g of dilution. Therefore, in 300 g of dilution, there will be (300 ÷ 5) × 1 = 60 g. **Answer:** d (60 g)

5.   A 1 in 10 dilution will contain 1 g of Dermovate Cream per 10 g of dilution. Therefore, in 200 g of dilution, there will be (200 ÷ 10) × 1 = 20 g. **Answer:** c (20 g)

6.   The cream contains 30% Arachis Oil BP. Therefore, there is 30 g of Arachis Oil BP per 100 g of cream. Therefore, in 50 g of final product, there is (30 ÷ 100) × 50 = 15 g. **Answer:** b (15 g)

7.   **Answer:** c (Freshly boiled and cooled purified water.)

8.   As the cream is for external use, the quantitative particulars are expressed per container. Although the quantitative particulars could be expressed as quantities per amount dispensed, for creams and ointments, it is customary to express the quantitative particulars as percentages. **Answer:** a (As percentages)

9. **Answer**: d ('Apply liberally as directed.')
10. **Answer**: b (Discard date of 4 weeks and 'For external use only'.)
11. Creams provide a suitable substrate for microbial growth which may cause spoilage or pathogenicity.
12. Collapsible tubes are the container of choice (*British Pharmacopoeia* 1988, p 650), although wide-necked jars can be used but there is a higher risk of contamination.
13. Dimeticone Cream BPC
    1. The product is used as a barrier cream (*Martindale* 35th edn, p 1849).
    2. This is an official preparation, therefore the formula is safe and suitable for use. As the cream is a barrier cream, using after bathing would be a suitable frequency.
    3. Prepare 30 g of Dimeticone Cream BPC.

Product formula
(from *British Pharmaceutical Codex* 1973, p 658)

	Master	500 g	50 g
Dimeticone 350 BP	100 g	50 g	5 g
Cetrimide BP	5 g	2.5 g	250 mg
Chlorocresol BP	1 g	500 mg	50 mg
Cetostearyl Alcohol BP	50 g	25 g	2.5 g
Liquid Paraffin BP	400 g	200 g	20 g
Freshly boiled and cooled purified water	444 g	222 g	22.2 g

4.
a. Dimeticone 350 BP is insoluble in water (*British Pharmacopoeia* 1988, p 200). Cetostearyl Alcohol BP is practically insoluble in water and, when melted, miscible with Liquid Paraffin (*British Pharmacopoeia* 1988, p 111). Cetrimide BP is soluble 1 in 2 parts water (*British Pharmacopoeia* 1988, p 111). Chlorocresol BP is slightly soluble in water and freely soluble in fixed oils (*British Pharmacopoeia* 1988, p 126). Liquid Paraffin BP is practically insoluble in water (*British Pharmacopoeia* 1988, p 415).
b. As creams are susceptible to microbial contamination, freshly boiled and cooled purified water will be used as the vehicle as per the product formula.
c. Chlorocresol BP is included in this product as the preservative as per the product formula.
d. Creams are for external use and so no flavouring is required.
   The following method would be used to prepare 50 g of Dimeticone Cream BPC from the formula above.
       Note that Cetostearyl Alcohol BP has a melting point of 49–56°C (*British Pharmacopoeia* 1988, p 111).
    1. Weigh 2.5 g Cetostearyl Alcohol BP on a Class II or electronic balance.
    2. Weigh 20 g Liquid Paraffin BP on a Class II or electronic balance.
    3. Weigh 5 g Dimeticone 350 BP on a Class II or electronic balance.

4. Weigh 22.2 g freshly boiled and cooled purified water on a Class II or electronic balance.
5. Weigh 250 mg of Cetrimide BP on a Class II or electronic balance.
6. Weigh 50 mg Chlorocresol BP on a Class I or sensitive electronic balance.
7. Melt the Cetostearyl Alcohol BP in an evaporating basin over a water bath to a temperature no higher than 60°C.
8. Add the Liquid Paraffin BP and the Dimeticone 350 BP to the melted Cetostearyl Alcohol BP and remove from the heat (this forms the *oily phase*).
9. Heat the freshly boiled and cooled purified water to 60°C.
10. Dissolve the Cetrimide BP and the Chlorocresol BP in the freshly boiled and cooled purified water and remove from the heat (this forms the *aqueous phase*).
11. When the oily phase and aqueous phase are both at about 60°C, add the oily (disperse phase) to the aqueous (continuous phase) with constant, not too vigorous stirring.
12. Stir until cool enough to pack. (Do not hasten cooling.)
13. Pack 30 g into a collapsible tube or amber glass jar, label and dispense.

5. A collapsible tube or plain amber jar would be most suitable.
6.
a. The product is official, therefore the following title would be suitable: 'Dimeticone Cream BPC'.
b. Quantitative particulars are not required as the product is official.
c. 'For external use only' will need to be added to the label as the product is a cream for external use.
d. 'To be used as a barrier cream after bathing.'
e. Not applicable.
f. The product is a cream and so will attract a 4-week discard date.
g.

Dimeticone Cream BPC	30 g
To be used as a barrier cream after bathing.	
Do not use after (4 weeks)	
For external use only	
Mrs Lily Evans	Date of dispensing

7. The patient would be advised to use the cream as a barrier cream after bathing. In addition, the discard date and fact that the product is for external use only would be highlighted to the patient.

14. Zinc oxide and calamine cream
1. Used as a barrier preparation for nappy rash and eczematous conditions and associated pruritus (*British National Formulary* 61st edn, p 706).
2. This is an unofficial preparation, therefore the formula will need to be checked to ensure that it is safe and suitable for use. The formula is comparable to official products containing the same ingredients listed in

the same/similar proportions (*British National Formulary* 61st edn, p 706). Therefore the product is safe and suitable for the intended purpose.

3. Prepare 20 g of zinc oxide and calamine cream.

Product formula

	Master	10 g	30 g
Zinc Oxide BP	15 g	1.5 g	4.5 g
Calamine BP	5 g	500 mg	1.5 g
Aqueous Cream BP	80 g	8 g	24 g

4.
a. Not applicable.
b. Aqueous Cream BP is used as the diluent in this preparation as per the product formula.
c. There is no preservative included as per the product formula.
d. Creams are for external use and so no flavouring is required.

The following method would be used to prepare 30 g of zinc oxide and calamine cream from the formula above:
1. Weigh 1.5 g Calamine BP on a Class II or electronic balance.
2. Transfer to a porcelain mortar.
3. Weigh 4.5 g Zinc Oxide BP on a Class II or electronic balance.
4. Add the Zinc Oxide BP to the Calamine BP and mix using the 'doubling-up' method.
5. Transfer the mixed powders on to a glass tile.
6. Weigh 24 g Aqueous Cream BP on a Class II or electronic balance.
7. Levigate the Aqueous Cream BP with the powder until smooth using the 'doubling-up' method.
8. Weigh 20 g and pack into a collapsible tube or amber glass jar. Label and dispense.

5. A collapsible tube or plain amber jar would be most suitable.
6.
a. The product is unofficial, therefore the following title would be suitable: 'Zinc oxide and calamine cream'.
b. Quantitative particulars are required as the product is unofficial. As the product is for external use, the quantitative particulars would be expressed by container (i.e. per 20 g):

Containing:
Zinc Oxide BP   15%
Calamine BP   5%
Aqueous Cream BP   ad 100%

c. 'For external use only' will need to be added to the label as the product is a cream for external use.
d. 'Apply THREE times a day as directed.'
e. Not applicable.
f. The product is a cream and so will attract a 4-week discard date.

g.

Zinc oxide and calamine cream	20 g
Apply THREE times a day as directed. Do not use after (4 weeks) For external use only	

Containing:

Zinc Oxide BP	15%
Calamine BP	5%
Aqueous Cream BP	ad 100%

Mrs Avril Asker   Date of dispensing

7. The patient would be advised to apply the cream three times a day as directed. In addition, the discard date and fact that the product is for external use only would be highlighted to the patient.

Chapter 6: Ointments, pastes and gels

1. A 1 in 5 dilution will contain 1 g of Hydrocortisone 1% Ointment BP per 5 g of dilution. Therefore, in 25 g of dilution, there will be $(25 \div 5) \times 1 = 5$ g. **Answer:** c (5 g)

2. A 25% concentration will contain 25 g of Salicylic Acid BP per 100 g of final product. Therefore, in 15 g of final product, there will be $(25 \div 100) \times 15 = 3.75$ g. **Answer:** c (3.75 g)

3. If there are 3.75 g of Salicylic Acid BP per 15 g of final product, there will be $15 - 3.75 = 11.25$ g of White Soft Paraffin BP. **Answer:** b (11.25 g)

4. A 0.75% concentration will contain 0.75 g (750 mg) of Salicylic Acid BP per 100 g of final product. Therefore, in 50 g of final product, there will be $(750 \div 100) \times 50 = 375$ mg. **Answer:** b (375 mg)

5. If there is 2.5 g of Hydrocortisone BP per 50 g of final product, there will be $(2.5 \div 50) \times 100 = 5$ g of Hydrocortisone BP per 100 g of final product. This is equivalent to 5% w/w. **Answer:** d (5%)

6. The prescription requests 2.5% Strong Coal Tar Solution BP. Therefore, there is 2.5 g of Strong Coal Tar Solution BP per 100 g of final product. Therefore in 20 g, there will be $(2.5 \div 100) \times 20 = 0.5$ g. **Answer:** b (0.5 g)

7. The prescription requests $(100 - 15 - 2.5 - 12.5 - 25)$% White Soft Paraffin BP. Therefore, there is 45 g of White Soft Paraffin BP per 100 g of final product. Therefore in 20 g, there will be $(45 \div 100) \times 20 = 9$ g. **Answer:** c (9 g)

8. **Answer:** c (Levigation)

9. **Answer**: d (3 months)
10. **Answer**: d ('Apply to the affected area.')
11.

a. Creams are semi-solid emulsions, either water-in-oil or oil-in-water. Ointments are also semi-solid, but are greasy. The base in which an ointment is made is immiscible with skin secretions, while the surfactants in creams allow better skin penetration

b. As creams are usually oil-in-water emulsions, they are easily miscible with skin secretions and have a 'vanishing' effect which is more acceptable to patients than the greasy messy alternative of an ointment. Creams are easier to apply and have a more pleasing texture.

12. Cetrimide Emulsifying Ointment BP

  1. Used as an antiseptic product for cleansing the skin (*Pharmaceutical Journal* 2004;273:351).
  2. This is an official preparation, therefore the formula is safe and suitable for use.
  3. Prepare 30 g of Cetrimide Emulsifying Ointment BP.

Product formula (from the *British Pharmacopoeia* 2007, p 2406)

	Master	100 g	10 g	40 g
Cetrimide BP	30 g	3 g	300 mg	1.2 g
Cetostearyl Alcohol BP	270 g	27 g	2.7 g	10.8 g
White Soft Paraffin BP	500 g	50 g	5 g	20 g
Liquid Paraffin BP	200 g	20 g	2 g	8 g

4.

a. Not applicable.
b. White Soft Paraffin BP is used as the diluent in this preparation as per the product formula.
c. There is no preservative included as per the product formula.
d. Ointments are for external use and so no flavouring is required.

  The following method would be used to prepare 40 g of Cetrimide Emulsifying Ointment BP from the formula above.
  Note that the melting points of the ingredients are: Cetostearyl Alcohol BP: 49–56°C (*British Pharmacopoeia* 1988, p 111).
  White/Yellow Soft Paraffin BP: 38–56°C (*British Pharmacopoeia* 1988, p 416).

  1. Weigh 20 g White Soft Paraffin BP on a Class II or electronic balance.
  2. Weigh 10.8 g Cetostearyl Alcohol BP on a Class II or electronic balance.
  3. Weigh 8 g Liquid Paraffin BP on a Class II or electronic balance.
  4. Melt together the White Soft Paraffin BP, Cetostearyl Alcohol BP and the Liquid Paraffin BP.
  5. Remove from the heat.
  6. Add the Cetrimide BP.
  7. Stir until cold.
  8. Weigh 20 g and pack into a collapsible tube or amber glass jar. Label and dispense.

5. A collapsible tube or plain amber jar would be most suitable.
6.
a. The product is official, therefore the following title would be suitable: 'Cetrimide Emulsifying Ointment BP'.
b. Quantitative particulars are not required as the product is official.
c. 'For external use only' will need to be added to the label as the product is an ointment for external use.
d. 'Use THREE times a day.'
e. Not applicable.
f. The product is an ointment and so will attract a 3-month discard date
g.

Cetrimide Emulsifying Ointment BP	30 g
Use THREE times a day.	
Do not use after (3 months)	
For external use only	
Mr Amarjit Singh	Date of dispensing

7. The patient would be advised to apply the ointment three times a day and use instead of soap. In addition, the discard date and fact that the product is for external use only would be highlighted to the patient.

13. Salicylic acid and sulphur ointment
1. To treat acne (*British National Formulary* 61st edn, p 726).
2. This is an unofficial preparation, therefore the formula will need to be checked to ensure that it is safe and suitable for use. Salicylic Acid BP is used in concentrations of 2–6% (*Martindale* 35th edn, p 1451) and Sulphur BP is used in concentrations of up to 10% (*Martindale* 35th edn, p 1452). Therefore the product is safe and suitable for the intended purpose.
3. Prepare 15 g of salicylic acid and sulphur ointment.

Product formula

	Master	10 g	20 g
Salicylic Acid BP	2 g	0.2 g	0.4 g
Precipitated Sulphur BP	3 g	0.3 g	0.6 g
Hydrous Ointment BP	95 g	9.5 g	19 g

4.
a. Not applicable.
b. Hydrous Ointment BP is used as the diluent in this preparation as per the product formula.
c. There is no preservative included as per the product formula.
d. Ointments are for external use and so no flavouring is required.

The following method would be used to prepare 20 g of salicylic acid and sulphur ointment from the formula above:

1. Weigh 400 mg Salicylic Acid BP on a Class II or electronic balance.
2. Transfer the Salicylic Acid BP to a glass mortar.
3. Weigh 600 mg Sulphur BP on a Class II or electronic balance.
4. Add to the Salicylic Acid BP in the mortar and grind together with the pestle to produce a fine evenly mixed powder.
5. Transfer the powder to a glass tile.
6. Weigh 19 g Hydrous Ointment BP on a Class II or electronic balance.
7. Transfer the Hydrous Ointment BP to the glass tile.
8. Triturate the powders with the Hydrous Ointment BP until a smooth product is formed.
9. Weigh 15 g of the product and pack in a collapsible tube or amber glass jar and label.

5. A collapsible tube or plain amber jar would be most suitable.

6.

a. The product is unofficial, therefore the following title would be suitable: 'Salicylic Acid and Sulphur Ointment'.

b. Quantitative particulars are required as the product is unofficial. As the product is for external use, the quantitative particulars will be expressed per container:

Containing:

Salicylic Acid BP	2%
Precipitated Sulphur BP	3%
Hydrous Ointment BP	to 100%

c. 'For external use only' will need to be added to the label as the product is an ointment for external use.

d. 'Apply to the patches TWICE a day.'

e. Not applicable.

f. The product is an ointment and so will attract a 3-month discard date.

g.

Salicylic Acid and Sulphur Ointment	**15 g**

Apply to the patches TWICE a day.
Do not use after (3 months)
For external use only

Containing:

Salicylic Acid BP	2%
Precipitated Sulphur BP	3%
Hydrous Ointment BP	to 100%
Mrs Helen Preston	Date of dispensing

7. The patient would be advised to apply the ointment to the patches twice a day. In addition, the discard date and fact that the product is for external use only would be highlighted to the patient.

Chapter 7: Suppositories and pessaries

1. **Answer:** b ('For rectal use only.')
2. **Answer:** d (3 months)
3. Zinc Oxide BP has a displacement value of 4.7. This means that 4.7 g of Zinc Oxide BP is required to displace 1 g of Hard Fat BP.

   Each suppository is required to contain 200 mg of Zinc Oxide BP. If 4.7 g of Zinc Oxide BP displaces 1 g of Hard Fat BP, 200 mg displaces $(1 \div 4.7) \times 0.2 = 0.0425$ g Hard Fat BP.

   Therefore, each suppository will contain $1 - 0.0425 = 0.9575$ g. **Answer:** d (957.5 mg)
4. Each suppository requires 60 mg of Phenobarbital BP. Phenobarbital BP has a displacement value of 1.1. This means that 1.1 g of Phenobarbital BP is required to displace 1 g of Hard Fat BP.

   Therefore, 60 mg of Phenobarbital BP will displace $(1 \div 1.1) \times 0.06 = 0.055$ g.
   Therefore, each suppository will require $1 - 0.055 = 0.945$ g (945 mg). **Answer:** c (945 mg)
5. **Answer:** d (Stemetil)
6. **Answer:** c (A suppository base should melt at just below 37°C.)
7. The ideal suppository (or pessary) base should (*Pharmaceutical Compounding and Dispensing*, p 77):
- Either melt after insertion into the body or dissolve in (and mix with) any rectal or vaginal fluid.
- Melt at body temperature. (It is preferable that the melting range should be slightly lower than 37°C as body temperature may be as low as 36°C at night.)
- Be non-toxic and non-irritant.
- Be compatible with any medicament and release it readily.
- Be easily moulded and removed from the mould.
- Be stable to heating above the melting point.
- Be stable on storage and resistant to handling.
8. Paracetamol 180 mg suppositories
    1. Pain relief (*British National Formulary* 61st edn, p 259).
    2. This is an unofficial preparation, therefore the formula will need to be checked to ensure that it is safe and suitable for use. The dose for a child (1–5 years) for paracetamol suppositories is 125–250 mg up to four times a day (*British National Formulary* 61st edn, p 259). Therefore the strength requested is safe and suitable for the intended purpose.
    3. Prepare 4 paracetamol 180 mg suppositories.

Calculations

Four suppositories are required; however, an overage will be needed to prepare this quantity successfully. Calculations are therefore based on the amounts required to prepare 10 suppositories.

Formula

	For 1 suppository	For 10 suppositories
Paracetamol BP	180 mg	1800 mg (1.8 g)
Hard Fat BP	to fill 1 × 1 g mould	to fill 10 × 1 g mould

Displacement value of paracetamol is 1.5.

1.5 g Paracetamol BP displaces	1 g of Hard Fat BP
Therefore 1.0 g Paracetamol BP displaces	$\dfrac{1.0 \text{ g of Hard Fat BP}}{1.5}$
Therefore 1.8 g Paracetamol BP displaces	$\dfrac{1.8 \text{ g of Hard Fat BP}}{1.5}$
	= 1.2 g Hard Fat BP.

Therefore the amount of Hard Fat BP required = 10 − 1.2 = 8.8 g

Product Formula

	10 suppositories
Paracetamol BP	1.8 g
Hard Fat BP	8.8 g

4.
a. Not applicable.
b. Hard Fat BP is being used as the base for this preparation.
c. There is no preservative included as per the product formula.
d. Suppositories are for rectal use and so no flavouring is required.

The following method would be used to prepare paracetamol 180 mg suppositories from the formula above.
Note that the melting point of Hard Fat BP is 30–45°C (*Martindale* 35th edn, p 1847):
1. Weigh 8.8 g Hard Fat BP on a Class II or electronic balance.
2. Transfer to an evaporating basin and melt over a water bath.
3. Weigh 1.8 g Paracetamol BP on a Class II or electronic balance.
4. Transfer to a glass mortar and grind to reduce particle size.
5. Levigate the Paracetamol BP with a small amount of the molten base on a glass tile.
6. Return to the remainder of the molten base and stir to mix well.
7. Stir until almost set and then pour into a disposable suppository mould and allow to set.
8. Trim the top and seal with the lid.
9. Transfer to a cardboard box and label.

5. Once manufactured, the suppositories should be individually wrapped in foil and placed in an ointment jar. Alternatively, if the suppositories were manufactured in a disposable mould, this could be labelled and dispensed directly to the patient.

6.

a. The product is unofficial, therefore the following title would be suitable: 'Paracetamol Suppositories 180 mg'.

b. Quantitative particulars are required as the product is unofficial. As the products are suppositories, the quantitative particulars will be expressed per suppository:

Each suppository contains:

Paracetamol BP	180 mg
Hard Fat BP	880 mg

c. 'For rectal use only' will need to be added to the label as the products are suppositories for rectal use.

d. 'Insert ONE into the rectum FOUR times a day when required.'

e. The *British National Formulary* (61st edn, p 260) recommends the following caution:

Label 30 – 'Contains paracetamol. Do not take anything else containing paracetamol while taking this medicine.'

However, as the products are suppositories, it would make sense to substitute 'take' with 'use'.

f. The products are suppositories and so will attract a 3-month discard date.

g.

Paracetamol Suppositories 180 mg   **4**

Insert ONE into the rectum FOUR times a day when required.

Do not use after (3 months)

For rectal use only

Contains paracetamol. Do not take anything else containing paracetamol while using this medicine.

Each suppository contains:

Paracetamol BP	180 mg
Hard Fat BP	880 mg

Miss Jessica Ramsden   Date of dispensing

---

7. The parent/guardian would be advised to insert one suppository into the patient's rectum four times a day when required. In addition, the discard date, the fact that the product is for rectal use only and the additional *British National Formulary* warning would be highlighted.

9. Metronidazole 170 mg suppositories

1. Used to treat infections (*British National Formulary* 61st edn, p 367).

2. This is an unofficial preparation, therefore the formula will need to be checked to ensure that it is safe and suitable for use. The usual rectal dose for a child (1–5 years) is 250 mg every 8 hours for 3 days, then every 12 hours (*British National Formulary* 61st edn, p 367). You are informed that the patient has been discharged from hospital on this regimen and requires 2 days' supply to finish the course. Therefore the preparation is safe and suitable for use.

3. Prepare six metronidazole 170 mg suppositories.

Calculations

Six suppositories are required; however an overage will be needed to prepare this quantity successfully. Calculations are therefore based on the amounts required to prepare 10 suppositories.

Formula

	For 1 suppository	For 10 suppositories
Metronidazole BP	170 mg	1700 mg (1.7 g)
Hard Fat BP	to fill 1 × 1 g mould	to fill 10 × 1 g mould

Displacement value of metronidazole is 1.7.
1.7 g Metronidazole BP displaces 1 g of Hard Fat BP. Therefore the amount of Hard Fat BP required = 10 − 1 = 9 g

Product formula

	10 suppositories
Metronidazole BP	1.7 g
Hard Fat BP	9 g

4.
a. Not applicable.
b. Hard Fat BP is being used as the base for this preparation.
c. There is no preservative included as per the product formula.
d. Suppositories are for rectal use and so no flavouring is required.

The following method would be used to prepare metronidazole 170 mg suppositories from the formula above. Note that the melting point of Hard Fat BP is 30–45°C (*Martindale* 35th edn, p 1847).
1. Weigh 9 g Hard Fat BP on a Class II balance.
2. Transfer to an evaporating basin and melt over a water bath.
3. Weigh 1.7 g Metronidazole BP.
4. Transfer to a glass mortar and grind to reduce particle size.
5. Levigate the Metronidazole BP with a small amount of the molten base on a glass tile.
6. Return to the remainder of the molten base and stir to mix well.
7. Stir until almost set and then pour into a clean, dry, matched suppository mould and allow to set.
8. Trim the tops and remove from the mould.
9. Wrap individually in foil.
10. Transfer to an amber glass jar and label.

5. Once manufactured, the suppositories should be individually wrapped in foil and placed in an ointment jar. Alternatively, the suppositories could be made in a disposable mould, which can be labelled and dispensed directly to the patient.

6.

a.   The product is unofficial, therefore the following title would be suitable: 'Metronidazole Suppositories 170 mg'.

b.   Quantitative particulars are required as the product is unofficial. As the products are suppositories, the quantitative particulars will be expressed per suppository:

Each suppository contains:

Metronidazole BP	170 mg
Hard Fat BP	900 mg

c.   'For rectal use only' will need to be added to the label as the products are suppositories for rectal use.

d.   'Insert ONE into the rectum THREE times a day.'

e.   The *British National Formulary* (61st edn, p 367) recommends the following cautions:

Label 4 – 'Warning: Do not drink alcohol while taking this medicine.'

Label 9 – 'Space the doses evenly throughout the day. Keep taking this medicine until the course is finished, unless you are told to stop.'

However, as the products are suppositories, it would make sense to substitute 'taking' with 'using' for Label 9.

f.   The products are suppositories and so will attract a 3-month discard date.

g.

Metronidazole Suppositories 170 mg   6

Insert ONE into the rectum THREE times a day.

Do not use after (3 months)

For rectal use only

Warning: Do not drink alcohol while using this medicine. Space the doses evenly throughout the day. Keep using this medicine until the course is finished unless you are told to stop.

Use at regular intervals. Complete the prescribed course unless otherwise directed

Each suppository contains:

Metronidazole BP	170 mg
Hard Fat BP	900 mg

Master Samuel Bridges   Date of dispensing

---

7.   The parent/guardian would be advised to insert one suppository into the patient's rectum three times a day. In addition, the discard date, the fact that the product is for rectal use only and the additional *British National Formulary* warnings would be highlighted.

Chapter 8: Powders and capsules

1.   **Answer:** a ('Store in a dry place.')

2.   **Answer:** c (*British National Formulary*)

3.   **Answer:** b (Storage conditions)

4.   The prescription requests one capsule to be taken three times a day for 5 days. Therefore, you will need to dispense $(1 \times 3 \times 5) = 15$ capsules. **Answer:** c (15)

5. The prescription requests two capsules to be taken every 4 hours for 1 week. Therefore, you will need to dispense (2 × 6 × 7) = 84 capsules. **Answer:** d (84)

6. The prescription is for 56 days' supply. For the first 14 days, the patient is to take 1 capsule three times a day. This is 14 × 3 = 42.
   For the remaining time (56 − 14 = 42 days), the patient is to take 1 capsule four times a day. This is 42 × 4 = 168.
   Therefore, you need to dispense 168 + 42 = 210. **Answer:** d (210)

7. **Answer:** c (To mix)

8. **Answer:** d (Size 3)

9. **Answer:** c (Dusting powders must be labelled 'Not to be taken'.) They should be labelled 'For external use only'.

10. 10 mg × 10 = 100 mg, which is below the accepted minimum weighable quantity of a Class II balance, which is 150 mg. All the other quantities, when multiplied by 10, are greater than 150 mg. **Answer:** a (10 mg)

11. A mix for 10 powders will contain 3 × 10 = 30 mg Ephedrine BP.
    This will be contained in a 200 mg portion of mix X. Therefore, the proportions of ingredients in mix X are:

    Ephedrine BP   30 mg

    Lactose BP   170 mg

    Multiplying both by 5 will produce weighable quantities:

    Ephedrine BP   30 mg × 5 = 150 mg

    Lactose BP   170 mg × 5 = 850 mg

    Answer: b (Ephedrine BP 150 mg Lactose BP 850 mg)

12. **Answer:** d (Verapamil BP 20 mg Lactose BP 180 mg)

13. Bendroflumethiazide 2.5 mg powders
    1. The product is used to treat oedema and hypertension (*British National Formulary* 61st edn, p 84).
    2. This is an unofficial preparation, therefore the formula will need to be checked to ensure that it is safe and suitable for use. Bendroflumethiazide is used at a dose of 5–10 mg in the morning daily or on alternate days; maintenance 5–10 mg 1–3 times weekly for oedema and 2.5 mg in the morning for hypertension (higher doses rarely necessary) (*British National Formulary* 61st edn, p 84). Lactose BP is a suitable diluent unless the patient is lactose-intolerant. Therefore the preparation is safe and suitable for the intended purpose.
    3. Prepare 5 bendroflumethiazide 2.5 mg powders.

Calculations

	One powder	10 powders
Bendroflumethiazide BP	2.5 mg	25 mg
Lactose BP	to 200 mg	to 2000 mg

We need to make a concentrated powder where every 200 mg of this concentrate (mix X) contains 25 mg of Bendroflumethiazide BP (mix X = 25 mg/200 mg).

We cannot weigh 25 mg. 150 mg is the minimum weight. Therefore to make mix X the same concentration we must multiply both parts of the fraction by the same number, i.e.:

8 × 25 mg = 200 mg

8 × 200 mg = 1600 mg

Mix X must therefore contain 200 mg/1600 mg. As we must weigh exact quantities, the formula for mix X is:

Bendroflumethiazide BP 200 mg

Lactose BP to 1600 mg (i.e. 1400 mg)

Now each 200 mg of mix X contains 25 mg of Bendroflumethiazide BP, therefore in our original formula we can substitute 200 mg of mix X for 25 mg of Bendroflumethiazide BP.

Formula for mix Y

Mix X	200 mg (containing 25 mg Bendroflumethiazide BP)
Lactose BP	1800 mg (to 2000 mg)

4.

a. Not applicable.

b. Lactose BP is used as a diluent (so long as the patient is not lactose-intolerant).

c. No preservative is included in the preparation.

d. Oral powders are swallowed with a draught of water and, as such, do not require flavouring.

The following method would be used to prepare 5 bendroflumethiazide 2.5 mg powders from the formula above:

1. Weigh 200 mg Bendroflumethiazide BP on a Class II or electronic balance.
2. Transfer to a porcelain mortar.
3. Weigh 1400 mg Lactose BP on a Class II or electronic balance.
4. Add the Lactose BP to the Bendroflumethiazide BP in the mortar using the 'doubling-up' method.
5. This is mix X.
6. Weigh 200 mg mix X and transfer to a clean porcelain mortar.
7. Weigh 1800 mg Lactose BP on a Class II or electronic balance.
8. Add the Lactose BP to mix X in the mortar using the 'doubling-up' method.
9. This is mix Y.
10. Weigh 200 mg aliquots of mix Y and wrap as individual powders.
11. Pack powders flap to flap and fasten together with a rubber band.
12. Pack into a cardboard box and label.

5. Once manufactured, the powders should be packaged flap to flap and enclosed with a rubber band. They can then be placed in a cardboard carton.

6.

a. The product is unofficial, therefore the following title would be suitable: 'Bendroflumethiazide Powders 2.5 mg

b. Quantitative particulars are required as the product is unofficial. As the products are powders for internal administration, the quantitative particulars will be expressed per powder:

Each powder contains:

Bendroflumethiazide BP	2.5 mg
Lactose BP	197.5 mg

c. 'Store in a dry place' will need to be added to the label as the products are individual dose powders.
d. 'Take the contents of ONE powder each morning in water.'
e. Not applicable.
f. The products are individual unit dose powders and so will attract a 3-month discard date.

g.

Bendroflumethiazide Powders 2.5 mg   **5**

Take the contents of ONE powder each morning in water.

Discard after (3 months)

Store in a dry place

Each powder contains:

Bendroflumethiazide BP	2.5 mg
Lactose BP	197.5 mg

Mrs Grace Browning   Date of dispensing

---

7. The patient would be advised to take the contents of one powder each morning in water. In addition, the discard date and the need to store the product in a dry place would be highlighted to the patient.

14. Imipramine 50 mg capsules
1. The product is used to treat nocturnal enuresis in children (*British National Formulary* 61st edn, p 236).
2. This is an unofficial preparation, therefore the formula will need to be checked to ensure that it is safe and suitable for use. Imipramine is used to treat nocturnal enuresis (for children aged 8–11 years) at a dose of 25–50 mg at bedtime (*British National Formulary* 61st edn, p 236). Lactose BP is a suitable diluent unless the patient is lactose-intolerant. Therefore the preparation is safe and suitable for the intended purpose.
3. Prepare 6 imipramine 50 mg capsules.

Calculations

	1 capsule	10 capsules
Imipramine Hydrochloride BP	50 mg	500 mg
Lactose BP	150 mg	1500 mg

4.
a. Not applicable.
b. Lactose BP is used as a diluent (so long as the patient is not lactose-intolerant).

c.   No preservative is included in the preparation.
d.   Capsules are swallowed whole with a draught of water and, as such, do not require flavouring.

The following method would be used to prepare 6 imipramine 50 mg capsules from the formula above:

1.   Weigh 500 mg of Imipramine Hydrochloride BP on a Class II or electronic balance.
2.   Transfer to a porcelain mortar.
3.   Weigh 1500 mg of Lactose BP on a Class II or electronic balance.
4.   Add the Lactose BP to the Imipramine Hydrochloride BP in the mortar using the 'doubling-up' method.
5.   Weigh 200 mg aliquots of the mixture and fill 6 size 3 capsules.
6.   Pack into an amber tablet bottle and label.

5.   A tablet bottle with a child-resistant closure.

6.

a.   The product is unofficial, therefore the following title would be suitable: 'Imipramine Hydrochloride 50 mg Capsules'.
b.   Quantitative particulars are required as the product is unofficial. As the products are capsules for internal administration, the quantitative particulars will be expressed per capsule:

Each capsule contains:

Imipramine Hydrochloride BP	50 mg
Lactose BP	150 mg

c.   Not applicable.
d.   'Take ONE capsule at night.'
e.   **Label 2** – 'Warning: This medicine may make you sleepy. If this happens, do not drive or use tools or machines. Do not drink alcohol.'
f.   The products are capsules and so will attract a 3-month discard date.
g.

Imipramine Hydrochloride 50 mg Capsules   6

Take ONE capsule at night. Discard after (3 months)

Warning: This medicine may make you sleepy. If this happens, do not drive or use tools or machines. Do not drink alcohol.

Each capsule contains:

Imipramine Hydrochloride BP	50 mg
Lactose BP	150 mg

Miss Jessica Felsham   Date of dispensing

---

7.   The patient would be advised to take one capsule at night. In addition, the discard date and the additional *British National Formulary* warning would be highlighted.

Bibliography

British National Formulary, 51st edn (2006) London: BMJ Publishing Group and RPS Publishing.

British National Formulary, 61st edn (2011) London: BMJ Group and Pharmaceutical Press.

British Pharmaceutical Codex (1968) London: Pharmaceutical Press.

British Pharmaceutical Codex (1973) London: Pharmaceutical Press.

British Pharmacopoeia (1980) London: HMSO.

British Pharmacopoeia (1988) London: HMSO.

British Pharmacopoeia (2004) London: TSO.

British Pharmacopoeia (2007) London: TSO.

Marriott J F, Wilson K A, Langley C A, Belcher D (2010) *Pharmaceutical Compounding and Dispensing,* 2nd edn. London: Pharmaceutical Press.

Martindale. The Extra Pharmacopoeia, 26th edn. (1972) London: Pharmaceutical Press.

Martindale. The Extra Pharmacopoeia, 31st edn. (1996) London: Royal Pharmaceutical Society.

Martindale. The Complete Drug Reference, 33rd edn. (2002) London: Pharmaceutical Press.

Martindale. The Complete Drug Reference, 35th edn. (2007) London: Pharmaceutical Press.

Further Reading

For further details on the formulation of the different dosage forms encountered within this text, student compounders are directed to the following core text:

Marriott J F, Wilson K A, Langley C A, Belcher D (2010) *Pharmaceutical Compounding and Dispensing*, 2nd edn. London: Pharmaceutical Press.

This text also provides an introduction to the history of pharmaceutical formulation and a collection of extemporaneous formulae.

It is beyond the scope of both this book and *Pharmaceutical Compounding and Dispensing* to go into detail on the science behind the different formulations described in the various chapters of the books. The texts listed below will be of use to compounders who wish to learn more about the science behind the formulations.

Aulton M E, ed (2002) *Pharmaceutics – The Science of Dosage Form Design*, 2nd edn. Edinburgh: Churchill Livingstone.
Collett D M, Aulton M E, eds (1990) *Pharmaceutical Practice*. Edinburgh: Churchill Livingstone.
Florence A T, Attwood D (2011) *Physicochemical Principles of Pharmacy*, 5th edn. London: Pharmaceutical Press.
Florence A T, Siepmann J, eds (2009) *Modern Pharmaceutics*, 5th edn. New York: Informa Healthcare USA.
Ghosh T K, Jasti B R, eds (2005) *Theory and Practice of Contemporary Pharmaceutics*, 2nd edn. Florida: CRC Press.
Sinko P J (ed.) (2011) *Martin's Physical Pharmacy and Pharmaceutical Sciences: Physical, Chemical and Biopharmaceutical Principles in the Pharmaceutical Sciences*, 6th edn. Baltimore, MD: Lippincott, Williams & Wilkins.

In addition, the following text will be of use to compounders wishing to practise and understand further pharmaceutical calculations.

Rees J A, Smith I, Smith B (2010) *Introduction to Pharmaceutical Calculations*, 3rd edn. London: Pharmaceutical Press.

Finally, the following text would be of interest to those compounders wishing to find out more about the history behind the formulations.

Anderson S, ed (2005) *Making Medicines: A Brief History of Pharmacy and Pharmaceuticals*. London: Pharmaceutical Press.

Appendices

Appendix 1
Glossary of terms used in formulations

Application	A liquid or semi-liquid preparation intended for application to the skin.
Bougie – nasal	A solid dosage form intended for insertion into the nostril.
Bougie – urethral	A solid dosage form intended for insertion into the urethra.
Cachet	An oral preparation consisting of dry powder enclosed in a shell of rice paper.
Capsule	An oral preparation consisting of a medicament enclosed in a shell, usually of gelatin basis. Soft gelatin capsules are used to enclose liquids and hard capsules to enclose solids.
Cream	A semi-solid emulsion intended for application to the skin. The emulsion may be an oil-in-water emulsion (aqueous creams) or water-in-oil type (oily creams).
Douche	A liquid preparation intended for introduction into the vagina.
Douche – nasal	A liquid preparation intended for introduction into the nostril.
Draught	A liquid oral preparation of fairly small volume and usually consisting of one dose.
Drops	A liquid preparation in which the quantity to be used at any one time is so small that it is measured as a number of drops, e.g. in a small pipette. Drops may comprise an oral preparation, usually paediatric, or may be intended for introduction into the nose, ear or eye, where the title of the product is amended accordingly.
Dusting powder	A preparation consisting of one or more substances in fine powder intended for the application to intact skin.
Elixir	An aromatic liquid preparation including a high proportion of alcohol, glycerine, propylene glycol or other solvent, and intended for the oral administration of potent or nauseous medicaments, in a small dose volume.
Emulsion	As a preparation, this term is generally restricted to an oil-in-water preparation intended for internal use.
Enema	An aqueous or oily solution or suspension intended for rectal administration.
Gargle	An aqueous solution, usually in concentrated form, intended for the treatment of the membranous lining of the throat.
Granules	A dry preparation in which each granule consists of a mixture of the ingredients in the correct proportions.

Inhalation	A preparation in which the active principle is drawn into the respiratory tract by inhalation. The active principle may be vapour when it is obtained from a liquid preparation by volatilisation, or it may be a solid when a special appliance, often an aerosol, is needed.
Injection	A preparation intended for parenteral administration which may consist of an aqueous or non-aqueous solution or suspension.
Irrigation	A solution intended for introduction into body cavities or deep wounds. Includes nasal and vaginal douches.
Levigation	This is the term applied to the incorporation into the base of insoluble coarse powders. It is often termed 'wet grinding'. It is the process where the powder is rubbed down with either the molten base or semi-solid base. A considerable shearing force is applied to avoid a gritty product.
Linctus	A viscous liquid preparation, usually containing sucrose, which is administered in small dose volumes and which should be sipped and swallowed slowly without the addition of water.
Liniment	A liquid or semi-liquid intended for application to intact skin, usually with considerable friction produced by massaging with the hand.
Lotion	A liquid preparation intended for application to the skin without friction. Eye lotions are lotions intended for application to the eye.
Lozenge	A solid oral preparation consisting of medicaments incorporated in a flavoured base and intended to dissolve or disintegrate slowly in the mouth.
Mixture	Liquid oral preparation consisting of one or more medicaments dissolved, suspended or diffused in an aqueous vehicle.
Mouthwash	An aqueous solution, often in concentrated form, intended for local treatment of the membranous lining of the mouth and gums.
Ointment	A semi-solid preparation consisting of one or more medicaments dissolved or dispersed in a suitable base and intended for application to the skin.
Paint	A liquid preparation intended for application to the skin or mucous membranes.
Pastille	A solid oral preparation consisting of one or more medicaments in an inert base and intended to dissolve slowly in the mouth.
Pessary	A solid dosage form intended for insertion into the vagina for local treatment.
Pill	A solid oral dose form consisting of one or more medicaments incorporated in a spherical or ovoid mass.
Poultice	A thick pasty preparation intended for application to the skin while hot.

Powder		A preparation consisting of one or more components in fine powder. It may be in bulk form or individually wrapped quantities and is intended for oral administration.
Suppository		A solid dosage form intended for insertion into the rectum for local or systemic treatment.
Syrup		A liquid preparation containing a high proportion of sucrose or other sweetening agent.
Tablet		A solid oral dosage form where one or more medicaments are compressed and moulded into shape.
Trituration		This is the term applied to the incorporation, into the base, of finely divided insoluble powders or liquids. The powders are placed on the tile and the base is incorporated using the 'doubling-up' technique. Liquids are usually incorporated by placing a small amount of ointment base on a tile and making a 'well' in the centre. Small quantities of liquid are then added and mixed in. Take care not to form air pockets that contain liquid, which, if squeezed when using inappropriate mixing action, will spray fluid on the compounder and surrounding area.

Appendix 2
Abbreviations commonly used within pharmacy

aa.	ana	of each
a.c.	ante cibum	before food
ad/add	addendus	to be added (up to)
ad lib	ad libitum	as much as desired
alt	alternus	alternate
alt die	alterno die	every other day
amp	ampulla	ampoule
applic	applicetur	let it be applied
aq	aqua	water
aq ad	aquam ad	water up to
aur/aurist	auristillae	ear drops
BNF		*British National Formulary*
BP		*British Pharmacopoeia*
BPC		*British Pharmaceutical Codex*
bd/bid	bis in die	twice a day
c	cum	with
cap	capsula	capsule
cc	cum cibus	with food
co/comp	compositus	compound
collut	collutorium	mouthwash
conc	concentratus	concentrated
corp	corpori	to the body
crem	cremor	cream
d	dies	a day
dd	de die	daily

dil	dilutus	diluted
div	divide	divide
DPF		*Dental Practitioners' Formulary*
DT		*Drug Tariff*
EP		*European Pharmacopoeia*
et	et	and
ex aq	ex aqua	in water
ext	extractum	an extract
fort	fortis	strong
freq	frequenter	frequently
f/ft/fiat	fiat	let it be made
ft mist	fiat mistura	let a mixture be made
ft pulv	fiat pulvis	let a powder be made
garg	gargarisma	a gargle
gutt/guttae/gtt	guttae	drops
h	hora	at the hour
hs	hora somni	at the hour of sleep (bedtime)
ic	inter cibos	between meals
inf	infusum	infusion
inh		inhalation/inhaler
irrig	irrigatio	irrigation
lin	linimentum	liniment
liq	liquor	solution
lot	lotio	lotion
m/mane	mane	in the morning
md	more dicto	as directed
mdu	more dicto utendus	use as directed
mist	mistura	mixture
mitt/mitte	mitte	send (quantity to be given)
narist	naristillae	nose drops
n/nocte	nocte	at night
n et m	nocte maneque	night and morning
np	nomen proprium	the proper name
neb	nebula	spray
ocul	oculo	to (for) the eye
oculent/oc	oculentum	an eye ointment
od	omni die	every day
oh	omni hora	every hour
om	omni mane	every morning
on	omninocte	every night
paa	parti affectae applicandus	apply to the affected part
pc	post cibum	after food
PC		prescriber contacted
PNC		prescriber not contacted
po	per os	by mouth
pess	pessus	pessary
pig	pigmentum	a paint
ppt	praecipitatus	precipitated
pr	per rectum	rectally
prn	pro re nata	when required

pulv	pulvis	a powder
pv	per vagina	vaginally
qds/qid	quarter die	four times a day
qqh/q4h	quarta quaque hora	every 4 hours
qs	quantum sufficiat	sufficient
R	recipe	take
rep/rept	repetatur	let it be repeated
sig	signa	let it be labelled
solv	solve	dissolve
sos	si opus sit	when necessary
stat	statim	immediately
supp	suppositorium	suppository
syr	syrupus	syrup
tds/tid	ter in die	three times a day
tinct	tinctura	tincture
tuss urg	tussi urgente	when the cough is troublesome
ung	unguentum	ointment
ut dict/ud	ut dictum	as directed
vap	vapor	an inhalation

Appendix 3
Formulae contents

The following lists, in alphabetical order, official product formulae contained in this book. The source of each formula is included with its entry.

Index